Learning to Think And Act Environmentally

While There Is Still Time

ABOUT THE AUTHORS

Lester W. Milbrath is the Director of the Research Program in Environment and Society and Professor Emeritus of Political Science and Sociology at the State University of New York at Buffalo. He is the author of many books, including *Envisioning a Sustainable Society: Learning Our Way Out* and *Environmentalists: Vanguard for a New Society,* both published by SUNY Press.

Ravi Gulati has a background in Management, but even as he was doing his MBA, he saw that the corporate lifestyle didn't much suit him. After being footloose for a while, he realised that his childhook passion was and remains the Environment. He is presently assisting a Voluntary Organisation in a small hamlet in the hills of Garhwal, where he is looking after their Medicinal and Aromatic Plants Programme. He is also researching and writing for a global advocacy film on Iron Deficiency Anemia.

M.L. Dewan is a Soil Scientist who retired as Chief Regional Bureau for Asia and Pacific, FAO (of U.N.). He has long experience in mountain systems and has built a Centre at Village Ranichauri in District Tehri Garhwal, U.P. Dr. Dewan has authored "People's Participation in Himalayan Eco-System Development" and four other bookis in "Discover Himalayan Series" and has also written over 100 articles and papers on Soil and Water Conservation, Watershed Management and Promotion of self-reliance in Himalayan communities.

Kriti Kamra is a budding artist and poet. She was still in school when she did these illustrations. She is presently studying at National Institute of Design, Ahmedabad.

Learning to Think And Act Environmentally

While There Is Still Time

Lester W. Millbrath

Indian Edition by Ravi Gulati

Afterword by Dr. M. L. Dewan

Illustrations by Kriti Kamra

CONCEPT PUBLISHING COMPANY

NEW DELHI-110059

Published in U.S.A. in 1996 by
State University of New York Press, Albany

ISBN : 81-7022-772-0 (H.B.)
ISBN : 81-7022-773-9 (P.B.)

First Indian Edition 1999

Indian Edition Published and Printed by:
Ashok Kumar Mittal
Concept Publishing Company
A/15-16, Commercial Block, Mohan Garden
New Delhi-110059 (India)

Phones : 5648039, 5649024
Fax : 091-11-5648053

For our grandchildren

Lester T. Milbrath
M. L. Dewan

CONTENTS

LIST OF FIGURES

FOREWORD

David T. Suzuki

For years, environmentalists have recited the phrase "Think globally, act locally" as a guide to action. However, in my experience, the global perspective has the effect of rendering us helpless - the issues appear so immense that whatever we do as individuals will have little effect on a seemingly unstoppable juggernaut. But we mustn't allow our feelings of impotence to paralyze us from taking action.

We live with the contradiction that while our massive complex brain has conferred a remarkable capacity for abstraction, inventiveness, and curiosity, as biological creatures our senses are tuned and responsive to our immediate surroundings. That's why we can be deeply moved by the plight of a child stuck in a well while ignoring the fact that over 35,000 children die daily of malnutrition and related problems. We are riveted to the predicament of underfed animals pulling heavily overloaded carts while unmoved by the extinction of perhaps 50,000 species a year! Hawala and corruption become a national obsession while issues like global warming disappear from public concern.

Increasingly, the world has been fragmented by the explosive growth in information - we are bombarded by proliferating television channels, the Internet, talk shows, news programmes and magazines. Unfortunately, most of the "infoglut" is, in fact, "infojunk", which is difficult to distinguish from more meaningful information. Consequently, our knowledge is shattered into fragments of "information" that has been disconnected from historical and social context. When I demand the source of some "fact" related by a student's statement, I am often told "I saw it on TV" or "I read it somewhere". "Facts" have become legitimated merely by their existence in print or in the electronic media.

Harvard's great biologist E.O. Wilson once told me that if human beings suddenly vanished overnight, except for a few human-specific species of parasites, other life forms would hardly register the loss, although the condition of many would improve markedly. Yet if all insects disappeared, life on the planet as we know it would be totally transformed. This makes it all the more astonishing that such an unprepossessing creature would have achieved a position where we, one species out of maybe 30 million, now co-opt 40 percent of the net photosynthetic activity on the planet and are causing massive changes in the biophysical features of the earth.

The biggest crisis we face today is our inability to "see" the dimensions of the building catastrophe that has occurred with lightning speed. Most people alive now were born after 1950, a period of unprecedented and unsustainable **change**. Most of modern technology was born and has evolved in this century, so, for today's citizens, change is a normal feature of daily life. An ecological perspective that views our existence within a context of geological time lets us see that what is happening now is suicidal. In this century, the human population has more than tripled and could double again in another 40 years. Since 1900, we have gone from a rural species to inhabitants of cities while our consumptive demands have more than quadrupled.

Les Milbrath has written and lectured prodigiously on the ecological crisis and the need for humanity to change direction onto a more sustainable path. He is now an elder in the movement that was set in motion by Rachel Carson's seminal work 'Silent Spring'. In a period when elders have been increasingly marginalized in our society, we have never needed their perspective more.

Using the Socratic method that worked so powerfully in Daniel Quinn's 'Ishmael', Milbrath leads us through the myths and filters that obstruct our view. We must confront what I call "sacred truths", beliefs that are so deeply embedded in our society's accepted notions that we never question them even when they often create the crises we are trying to overcome. I'm speaking of such fallacies as our species' escape from the boundaries and limits of the natural world, faith in the capacity of science and technology to provide the knowledge and tools to understand and control our surroundings, and the belief that the human-created economy subsumes all else as a priority. Unless we confront the flaws in such beliefs and discard them, we will never be able to move seriously to a different path that is sustainable.

Milbrath's book is deceptively modest. It confronts inescapable facts of limits and our place as animals within the global biosphere. For all who care about what we are leaving for our children and grandchildren, this is an important contribution.

PREFACE

My neighbour is a cheerful, thoughtful man with a deep sense of social responsibility. A few years ago he said, "Some time I would like to sit down with you and learn more about the environment. Sometimes things seem to be going along OK, but other times they seem to be getting worse. I have this vague feeling that maybe we aren't taking sufficient precautions to make sure things don't go wrong in the future". He reflects the feelings of many Indians who are torn between the promise of development, and the increasing importance being given to environmental issues in the media.

Most adults, and children, get their information about the environment from the mass media. While presenting fragmented environmental information, the media also present a way of thinking about it. The current mode of environmental thinking is so deeply embedded in our culture that we are unaware that we may have been taught to think in a special way. Regrettably, contemporary public discourse uses flawed basic principles for environmental thinking. If we are to rescue our planet's life systems, and save our society, we must relearn our basic understanding of how the world works.

Many years of teaching environmental policy and politics to college students has shown me that before students can imagine long-term effective solutions to environmental problems they must first learn fundamental principles about how the world works. They come to class poorly prepared because our schools have been terribly deficient in providing environmental education. One reason environmental education is slighted is that most teachers believe that their curriculum is already so overloaded with mandated requirements that adding another topic seems unbearable. Another reason is that most teachers themselves are poorly prepared to teach about the environment.

My neighbour's desire to learn stimulated me to try to fulfill several needs for greater clarity about how the world works with one short, easy-to-read book. The basic principles of environmental thinking are not difficult. They should be taught in our schools because bright fifth or sixth class students can grasp most of the principles. Teachers need to learn them and should be sharing them with their students. Mass media professionals need to understand them better so they can present their stories accurately. Policy makers should know them to make wise policies. And every citizen needs to know them to make wise civic decisions. Knowing these principles is just as basic to being educated as is knowing history.

Hoping to make this book more readable, I have portrayed a conversation with my neighbour, who really wants to learn this new way of thinking. We invite you to sit in with us as we explore a more adequate way of understanding how the world works.

ACKNOWLEDGEMENTS

Attempting to make complex concepts understandable to the average person required seeking unusual amounts of feedback from a variety of people. I am especially indebted to George Besch, filmmaker and fishing buddy, for not only reading and criticizing each draft, as well as consulting on many questions of presentation, but also for rounding up his friends to read and criticize–most of them anonymously. Paul Reitan, professor of geology and longtime friend, read and criticized three drafts. Bill and Maureen Millbrath critically read two drafts and rounded up additional readers. Bill Wood, Sheldon Kamienecki, Helen Besch, Mike Mitchell, Mike Rinaldo, Teresa Milbrath, and Kirsten Milbrath read and criticized a draft. Lia Marcote came up with the idea for the subtitle of the book. Kate Miller, with a talented sharp eye, proofread the manuscript. Clay Morgan, acquisitions editor at SUNY Press, provided good advice and rounded up readers to provide additional valuable feedback. He also provided cogent advice while editing the final manuscript. David Suzuki took time from his feverish schedule to provide wise counsel and the foreword.

My understanding of these concepts has been stimulated by teaching classes in environmental politics and environmental sociology for 15 years. The queries of students provided frequent clues as to which concepts described in which ways are crucial for full understanding of this subject. Three classes at SUNY at Buffalo and two classes at the University of California at Irvine read drafts of this book and provided valuable feedback. Several students in a Future Society class offered advice for designing the book cover. My students are too numerous for acknowledgment here but each will know that they played a role in shaping this book. Also, I am indebted to a small army of environmental scientists whose research on a myriad of aspects of our complex environment continues to enlighten mankind. My thanks to them all.

Finally, I am grateful to Tom Toles and *The Buffalo News* for permission to reprint the Toles cartoons that appear throughout the book. These cartoons have been Indianized by Kriti Kamra.

Chapter 1

Introduction

My neighbour Madan says, "I HAVE THIS VAGUE UNEASINESS THAT MAY BE THINGS AREN'T SO GOOD WITH OUR ENVIRONMENT. YOU TEACH YOUNG PEOPLE ABOUT THE ENVIRONMENT, WHAT DO THEY THINK?"

Frankly, many of them are aware of the problems facing us. But their awareness is usually limited to their own surroundings. To most city people, environmental degradation is synonymous with air pollution, because our cities are so clogged with smoke. The most accused are diesel driven vehicles such as trucks and buses. Although 2 stroke engines of the popular 100 cc motorcycles also pollute, they are not so 'bad', according to popular conception, because their emissions are invisible.

For a lot of young people, protection of the Environment and catching up with the developed countries are conflicting objectives. The Environmentalists always seem to be opposing everything that represents prosperity. Nevertheless, realizing the importance of saving the Environment as they see it, they usually advocate a balance. A balance in which saving the environment means making more pollution control equipment, and making and enforcing pollution control laws, for which they look upon funds or resources as the vital input in short supply, especially for a poor country like India. Resources are generated by Western style development and so, especially after the break up of the Soviet Union, there is no alternative to Economic liberalisation and reform.

This optimism seems to be overlaid over a profound pessimism. Our institutions are unable to cope with the problems we face. Why? Corruption is found at the highest levels. Why have we no faith in the system? Why do we think of the illiterate, hungry masses as a massive group lurking out there to retard 'our' progress? Why are so many people out of work? Why are slums growing like cancer? Why is there so much terrorism and crime in the land of Gandhi?

THOSE SOUND LIKE SEPARATE PROBLEMS TO ME. ARE YOU SUGGESTING THAT THEY'RE CONNECTED?

In some respects these are separate problems, but they're similar in that **all of them are connected in one way or another to the faulty way we think about the environment**. By that I mean that many of the premises underlying the beliefs in our society about how the world works are not true. Our ways of thinking are taught to us by our culture as we're growing up. Of course, as we mature, all of us develop our own ways of looking at things. But the underlying premises that shape our thinking come from our culture; they are simply taken for granted as we communicate with others. In cities, the media, especially television and magazines, constantly reinforce those unspoken premises.

WHAT DO YOU MEAN? GIVE ME AN EXAMPLE.

OK! Last evening, the television newsreader said, "Good news tonight. The Commerce Department reported today that the Gross Domestic Product (GDP) was up 1.6% over last year, mainly due to increased sales of cars and petrol. Well, why did he say it was "good news"? Apparently, he assumed everyone would believe that a rise in the GDP is good news; he never dreamt that anyone would disagree. So indirectly, the viewer was taught by that statement that economic growth is good.

EVERYONE I KNOW THINKS ECONOMIC GROWTH IS GOOD. ARE YOU SUGGESTING IT ISN'T GOOD?

Your response illustrates my point. Our present mainstream thinking teaches us that economic growth is good, and you believe everyone agrees with that. Actually, some of the leading thinkers in the World are sure that it is not good.

WHY WOULD THEY SAY THAT? WHO ARE THESE THINKERS?

Before I can answer those questions to your satisfaction, I should share with you some basic principles that most people in our society don't grasp. If you can bear with me for a while, I'll show you this new way of thinking.

OK, I'M SKEPTICAL BUT I'LL GIVE IT A TRY. HOW DO WE BEGIN?

Chapter 2

Beliefs Empower and Deceive Us

First, it's important to understand that we all have pictures in our mind about how the world works that we use to organize our daily lives. We can't get through the day without a set of basic beliefs about the workings of the world. As I said earlier, mostly we use beliefs handed to us by our culture (that includes religion, education, science, folk wisdom, and, especially these days, ways of thinking portrayed by the mass media).

Usually we absorb these basic beliefs from our culture early in childhood in a process social scientists call "socialization". Parents, teachers, elders, and media socialize young people to approved ways of thinking by their comments, actions, and counseling. We so thoroughly accept our culture's beliefs about how the world works that we hardly ever think about them even though they underlie everything we think and do. For instance, most people hold the assumption that the world will work in the future much like it has in the past.

ARE YOU SUGGESTING THAT'S NOT TRUE?

Recent evidence strongly suggests that humans are changing the way the world works.

OH YES, I CAN SEE CERTAIN CHANGES LIKE NEW TECHNOLOGY. BUT DOESN'T THE

NATURAL WORLD WORK THE WAY IT ALWAYS HAS? ISN'T IT BASICALLY PREDICTABLE?

Actually, we're making it less predictable and we're seriously threatening the viability of the planet's life systems. We never set out to do that, of course. Ironically, we were only doing better and better what we have always been told is good and right. For example, folk wisdom says that we should live as long as possible and have children to carry on our family line. Humans have been so successful in doing both things that demographers now talk about a population explosion. Persons over 85 years of age are the most swiftly increasing age group.

HOW FAST IS THE POPULATION GROWING?

When humans first evolved, their numbers grew slowly. It took over 250,000 years to reach one billion people early in the nineteenth century. But the pace quickened and the second billion was added in 130 years. The third billion took only 30 years and the fourth only 15 years. At our current pace, the next billion will take a

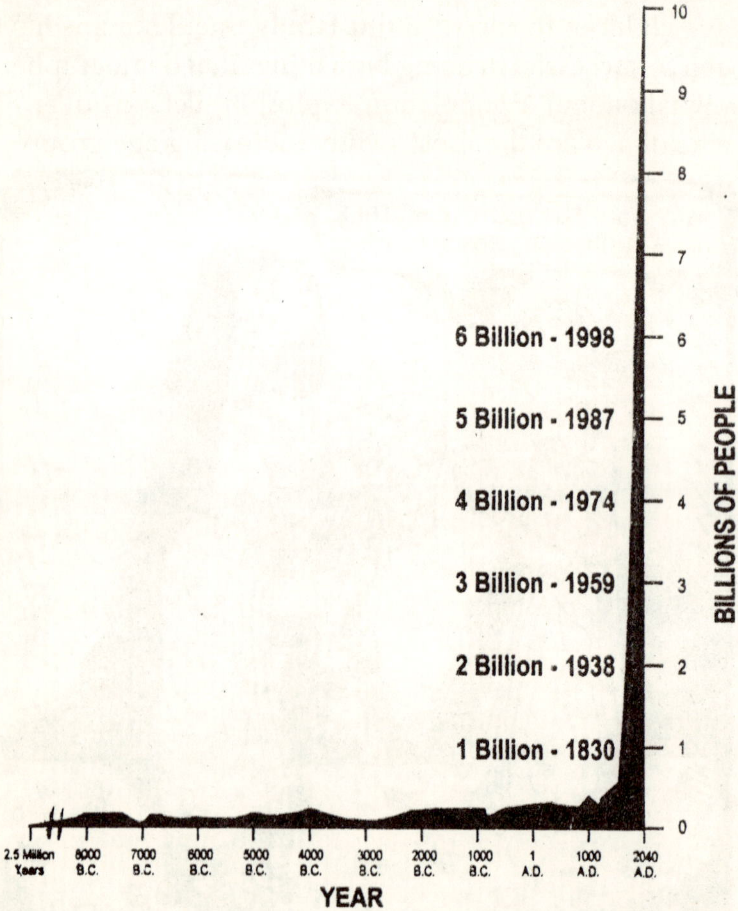

Figure 2.1: World Population Growth Throughout History and Near-Term Future

little over ten years. Every two days the **net growth** in humans is enough to fill a city the size of San Francisco in the U.S.. Take a look at this graph (see figure 2.1).

Notice how the line rises very slowly for many centuries and then rises steeply as the Industrial Revolution takes hold. Human population took 10,000 generations to grow to two billion; now, in one lifetime, it will grow another five billion. We currently add over 90 million people to the planet every year. That's about the size of Mexico or Germany. There are more people alive today than have ever died. If other creatures could speak, they would call us an epidemic. Current world population is close to six billion. If present trends continue, the population will double again to 12 billion by the year 2040. Theoretically, it could double again to 24 billion before the end of the next century.

WILL THAT REALLY HAPPEN?

Probably not. The United Nations held its third World Conference on Population and Development in Cairo in September 1994. It set a goal of slowing the rate of population growth so that the total world population will top off somewhere between seven and eight billion. I think it will grow larger before it can be topped off, but nature eventually may force us to reduce our numbers to less than seven billion. The Population Conference strongly urged all nations to slow down population growth as soon as possible. Women played a much larger role in the Cairo conference than in previous population conferences. The emphasis shifted from focusing on numbers to accenting the emancipation and empowerment of women, especially in providing education and health services to them. That is believed to be the most effective strategy for slowing population growth. If we fail to stop population growth by voluntary measures, nature will probably use catastrophes like famine or disease to prevent us from going much beyond seven or eight billion.

BUT WON'T BILLIONS MORE BABIES STILL BE BORN?

Yes, mainly because there are so many young people in the world who will soon be mothers and fathers. Conventional thinking also tells us that everyone should try to become prosperous and that high consumption makes for a healthy economy. So those additional people probably will scramble for money and consume as much as possible.

Chapter 3

The Hidden Danger in Doubling Times

WHAT WILL BILLIONS MORE PEOPLE TRYING TO LIVE HIGH-CONSUMPTION LIFESTYLES DO TO THE EARTH?

Environmentalists use the formula I=PAT to illustrate the impact of a given society on the earth. Environmental impact (I) is the function of its population (P), its capital stock per person or affluence (A), and the environmental consequences of the technology it uses (T). We can't measure those factors precisely enough to know if they are additive or multiplicative but we do know that each of the three factors contributes powerfully to the total impact.

If all the people we'll be adding to the earth try to be high consumers, we'd have to extract resources from the earth much faster and process them into products faster. Eventually, all that stuff will turn into waste, which we then throw back into nature. This processing and discarding of materials is called **throughput**. The current doubling time of 30 to 50 years for population could shorten to only 10 to 15 years for a doubling of total planetary impact. The kicker is that nature simply doesn't have enough resources for endless growth via throughput, and it cannot absorb all the wastes we would create.

Modern society constantly urges us to speed up throughput. Gross National Product (GNP) was the term economists used for many years to measure throughput. Because throughput creates so much waste, some wags have suggested it should be called Gross National Garbage. Lately, economists have been using the phrase Gross Domestic Product (GDP) to identify goods made in a given country even though the factory may be owned by a firm based in another country. Both GNP and GDP are measures of throughput.

SO, DESPITE THE CONVENTIONAL WISDOM THAT SAYS ECONOMIC GROWTH IS GOOD, YOU'RE CALLING FOR LIMITS TO GROWTH?

Yes, I am. It's the only sensible choice in today's world. The longer we delay making this choice, the worse the problem becomes.

IS THIS CRISIS OF GROWTH SOMETHING NEW IN HISTORY, OR HAS IT ALWAYS BEEN THIS WAY?

In some ways it's an old story. Even if a swiftly growing society exceeds the **carrying capacity** of its natural systems, the population keeps growing for a while. The excess population is called **overshoot**. Usually populations that go into overshoot injure natural systems severely, leading to a crash. That's called **dieback**. The post-crash carrying capacity typically is much lower than the capacity that could have been sustained had the population grown more slowly. There are numerous examples of overshoot, ecosystem destruction, and dieback in the past that most history books don't tell us about. Some tribes, even some civilizations, have overpopulated their ecosystems and devastated them to the point that they had to be abandoned. Some of those ecosystems recovered when humans left, but others turned into permanent desert.

In today's world, however, the huge mass of people has the capability to change the life systems of the **whole planet**. We no longer have the choice to move and occupy a new ecosystem if we destroy the one we're in.

BUT I ALWAYS THOUGHT NATURE WAS SO POWERFUL AND PERMANENT THAT WE COULDN'T AFFECT THE WAY IT WORKS NO MATTER WHAT WE DID. THE HOST ON A RADIO TALK SHOW JUST YESTERDAY CLAIMED THAT THERE'S NO EVIDENCE WE'RE HURTING NATURE.

They use the faulty thinking that got us into our environmental predicament. Their most deplorable mistake is that they focus on what has happened up to now and don't look ahead and ask what will happen to our ecosystems as people and throughput swiftly increase.

Another typical mistake antienvironmentalists make is that they don't examine the interconnected effects of a wide range of factors. For example, the mass of humans now being born will be enticed by powerful new technologies that business will urge people to buy and use. Business will resist any controls on those technologies prior to their release. Yet, some of those powerful new technologies will do unforeseen, but serious, damage to the viability of life systems. With billions more people using powerful technologies, the damage may not be contained before millions of humans and other creatures have died.

Paradoxically, we humans are the most adaptable, the most powerful, the most successful species to ever live on this planet. Yet if we refuse to look ahead and refuse to limit our numbers while still trying to consume more and more, we'll injure life systems and reduce their ability to support life at the very time all those new people will be seeking life support.

So at this stage of history, trying to achieve "success" in the conventional sense will make us **victims of our own success**. Other creatures will also become victims. We never set out to change life systems, but that's what is happening as a result of following conventional wisdom.

SURELY WE CAN LOOK AHEAD AND MAKE CHANGES TO PREVENT A CRISIS?

I'm not very confident we will. So far, mankind has strongly favored looking only at the near-term future and has refused to take seriously what could happen in the long term. We also have not yet learned to **think in doubling times**.

YOU USED THAT TERM BEFORE. WHAT EXACTLY DO YOU MEAN BY THINKING IN DOUBLING TIMES?

Here's a riddle that illustrates it. If a water lily, growing in a pond, doubles in size every day, and covers the pond on the 30th day, on what day did it cover half of the pond? The intuitive answer that first pops into our mind is the 15th day, but on closer reflection we realize that the lily would not have covered half the pond until the 29th day; it requires only one more doubling to cover the whole pond. One day's growth equals all the growth of the previous twenty-nine. **When a base is already large,**

another doubling can be enormous. Growth that used to be welcome can suddenly become a great threat. Consequently, many environmentalists are worried that the world is at the "29th day", because the growth in human population and in our impact on the planet over the next 50 years will very likely exceed all the growth since humans first came to live on the planet.

Population doubling times vary considerably from region to region in today's world. At current rates, population will double in Africa in only 24 years, 37 years in South America, 65 years in East Asia, 98 years in North America, and 1,025 years in Europe.

In India, population will double in 32 years.

DO YOU THINK THE WORLD'S LEADERS REALIZE WHAT'S HAPPENING?

Some do, but most don't. You see, our minds are locked up in our old theory about how the world works. Here in India most people are convinced that the American way works quite well. The U.S.A. is an economic superpower and a land of opportunity. It is seen as the immigrant's paradise. To get a green card is to have arrived. With that mindset we fail to recognize how even their world has changed.

The wise people we look up to, great philosophers, saints or prophets, for example, never confronted the kind of world we confront today. Yet in their wisdom they could foresee what could happen if everybody became acquisitive, and consumed as much as they could. Our culture teaches us to live simply not only because that is vital for our spiritual development [*Aparigraha* (Non-Accumulation) is an important vow for the spiritually inclined], but also because it is wise and good to live lightly upon the Earth.

Somewhere along the line, the balance of the extreme contradiction between unrestrained materialism and unrelenting asceticism that is so typically Indian began to tilt in favour of material progress. In any case, beliefs that were wise for the past may not be wise for the future. We need to relearn how the world works as revealed by environmental science and develop ways of thinking appropriate to the reality of that new world.

BUT WE'VE BEEN TAUGHT THAT THE WISDOM OF THE GREAT THINKERS IS GOOD FOR ALL TIME.

The great thinkers of the past held beliefs appropriate to the world they knew. But when we borrow from the past, we do it selectively. Our colonial history and the resultant loss of self respect and dignity led us to confuse modernization with Westernization, a tendency that still largely persists. Wisdom bifurcated into the ancient home-grown variety, meant to be preached, and the relatively new and 'successful' variety, meant to be practised. So, while we like to imitate the American way, we rarely admit that without imparting our justifications an Indian flavour.

Our world is now so different that our predominant way of thinking is leading us to damage global life systems so seriously that our actions could lead to the destruction of our civilization. **Our only way out is to change the way we think.**

I SENSE THAT THINGS AREN'T QUITE RIGHT, BUT THEY SURE DON'T SEEM TO BE AS BAD AS YOU SUGGEST. JUST THE OTHER DAY I

READ THAT ENVIRONMENTAL LAWS WERE WORKING WELL AND THAT WE SHOULD STOP TALKING ABOUT DOOMSDAY.

I read that article too and I agree with the author that pollution control laws in the developed world are slowly becoming effective. If we hadn't simultaneously grown so swiftly in population and economic activity, the positive results would be even more visible. The situation in the Third World, however, is swiftly getting much worse because population is growing much faster there and environmental protection is very weak. That article also focussed on pollution and does not address many of the other environmental problems that we will be discussing. This is no time for complacency because I believe our problems will get much worse if we don't change the way we think and act. I perceive a future very different from the future our national leaders portray. To explain why I see a different future, I'll try to show you the forces that are driving global systems and explain how human activities contribute to those forces.

Chapter 4

Learning to Think Systemically

Most of us are taught to think **linearly**; that is, our thinking goes in straight lines. Time runs in a line from yesterday to today to tomorrow. A causes B causes C causes D; we're taught to look for cause and effect. Mechanical thinking is mostly linear: find the cause, fix it, and get the desired effect. Linear, mechanical-style thinking plays a large role in public policy. If a crime has been committed, find the culprit, toss him in the prison, and throw away the key - problem solved. The article we discussed that claims our environmental problems are solved used linear thinking.

The natural world is not so simple; it is a huge system of systems. A fundamental principle of environmental thinking is that **biospheric elements circulate in systems**. For example, environmental scientists talk about the carbon cycle or the nitrogen cycle or the water cycle as they admonish us to **think systemically**. And be sure to note I didn't say systematically; I speak of systems thinking. Systems thinkers focus on wholes rather than on parts. Within wholes they concern themselves with relationships more than objects, with processes more than structures, with networks more than hierarchies. In a system, a given effect not only radiates through the system, it also feeds back and changes the factor that caused it.

I THINK I UNDERSTAND, BUT GIVE ME AN EXAMPLE.

I'll use the carbon cycle as an example. Carbon dioxide, CO_2, makes up only about 0.035 percent of the earth's atmosphere, but it's vital to living things. Plants use chlorophyll (that's what gives them their green color) to take energy from the sun, carbon from the CO_2 in the air, and water containing nutrients from their roots to make $C_6H_{12}O_6$ (glucose) as a basic building block in new structures. This creative process is called **photosynthesis**. As the plants use the carbon atoms as a building block in their growth, they simultaneously release oxygen molecules. The new plant structures (leaves, stems, seeds, fruit) contain bound-up energy. The solid trunk of a tree is mostly carbon, which is why it burns so well.

Animals inhale the oxygen released by the plants and use it to burn (digest) food - which they also get, directly or indirectly, from plants - thus recombining carbon and oxygen into CO_2. When they exhale the CO_2, it circulates back to the plants, and the plants use it to make more food. The carbon circulates in a system from plants to animals and back again. Each part of the system is dependent on the other parts. A change in one aspect of a system usually reverberates through the whole system.

THAT SEEMS CLEAR, BUT ANOTHER EXAMPLE OF SYSTEMIC THINKING WOULD HELP.

A litre of petrol contains a lot of carbon. When it's burned in your car, the carbon combines with oxygen to form CO_2. The carbon in oil does not come from plants living now but from plants that lived millions of years ago.

So burning millions of litres of petrol today quickly releases extraordinary amounts of CO_2 that nature had not figured into the current carbon balance between plants and animals. But now there aren't enough growing plants to take up the extra CO_2 so it's accumulating rapidly.

In fact, we humans further insult the system by cutting down forests to get timber or to clear land for farming or building factories and cities. With less forest to take up CO_2, it builds up even faster. Now environmentalists, most atmospheric scientists, and some political leaders are warning us that the extra carbon dioxide is adding to the greenhouse effect that warms the planet and probably will change the climate system.

HOW WILL IT CHANGE THE CLIMATE SYSTEM?

Let's postpone that question until I introduce the next major principle. The point here is that different parts of a system are so interconnected that **we can never do merely one thing**. We should always ask, "**And then what?**"

A MINUTE AGO YOU SAID, "AN EFFECT FEEDS BACK". WHAT DO YOU MEAN?

In a system, each action creates an effect that influences other parts of the system. It often alters the system sufficiently to change the thing that started the action. Let me diagram it for you.

If A causes B, and B causes C, C may feed back and change A. When that happens, we call it a "feedback loop".

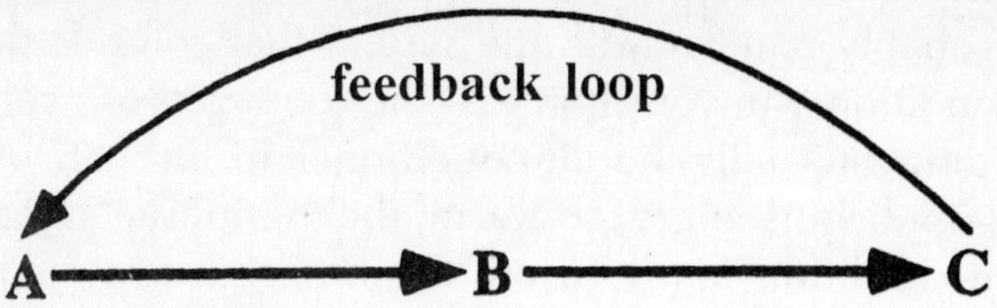

Figure : 4.1 :Feedback

Here is a simple example from nature. Imagine a forest ecosystem with foxes and rabbits living in it. Of course it would have many other animals living there as well, but we'll concentrate on rabbits and foxes. The rabbits eat plants for food, and the foxes eat rabbits for their food. So, if the foxes eat well and increase their population, that will inevitably decrease the population of rabbits. Since foxes don't eat plants, some of them will then starve. With fewer predator foxes, the rabbit population increases again and, later, the fox population can also increase when they have more rabbits for food. That's generally how nature keeps any one species from dominating and destroying the good functioning of a system.

SO THOSE INCREASES AND DECREASES WERE FEEDBACKS?

Yes. The feedback the foxes got when they ate too many rabbits came in the form of many foxes dying. When a feedback causes a situation to return toward its original position, it is called a **negative feedback loop**. Growth or decline in either population influences what the other can do. Since negative feedbacks correct and control a system they can be called **corrective feedbacks**.

Corrective (negative) feedback loops are usually built

into mechanical systems as well. For instance, you're probably familiar with the system that controls the air-conditioner in your home. If the room gets too cold, the thermostat tells the air-conditioner to turn off. As the room begins to get too warm, the thermostat signals the air-conditioner to turn on. So the corrective feedbacks built into that system keep returning the system to its original setting and are essential to keep it in control.

WHAT HAPPENS WHEN A SYSTEM GOES OUT OF CONTROL?

Systems that go out of control are likely to crash. Typically, a **positive feedback loop** is involved when a system goes out of control. When a system has a positive feedback, each factor stimulates the other to increase. There aren't very many examples of them in nature, because they don't last long. But have you ever heard a loudspeaker screech?

SURE, IT STARTLES PEOPLE AND THEY COVER THEIR EARS.

That happens because a positive feedback loop is at work. When someone turns up the amplifier in a loudspeaker, the speakers get louder. The louder sound reaches the microphone and has an effect similar to shouting into it. The higher signal from the microphone tells the amplifier to make the speakers even louder, which blasts increasingly louder into the microphone. The positive feedback quickly accelerates the system to the point of breakdown. If someone doesn't turn down the amplifier quickly, or cover the microphone, the system crashes. For that reason we can call them **runaway feedbacks**.

ARE THERE SYSTEMS WITH POSITIVE FEEDBACK LOOPS IN NATURE?

Sometimes an organism gets into a situation where there's plenty of food and no competitors or predators to act as a negative feedback, allowing it to reproduce epidemically. The population literally explodes. An example would be an algal bloom. Some algae bloom so quickly they're called a "killer tide". Explosive growth of the algae creates terrific overcrowding, and nutrients are quickly exhausted. I was in Norway a few years ago when a killer algal tide swept along the coast. It advanced 50 miles a day and killed almost every living thing in its path.

Recent research showed that a type of killer algae (pfiesteria piscimorte) flourishes when water is rich in phosphates, and that it spontaneously transforms from plant to animal and back again. In animal form (giant amoeba) it consumes the bodies of the fish it kills, allowing it to reproduce even more explosively. After killing everything within reach, the tide subsides. The ecosystem where this event occurs is usually devastated. That's why runaway (positive) feedbacks are unsustainable and should be avoided.

Most runaway feedbacks are carelessly caused by humans who don't anticipate the second, third, and fourth order consequences of their actions. The killer algal tide in Norway got its rich phosphate nutrients from fertilizer runoff from farms; the farmers only wanted to make money when they spread the fertilizer and never thought about stimulating a killer algal tide.

Many human actions inadvertently create runaway feedbacks. For example, on a hot day people are likely to turn on their air conditioners. The heat from inside the house or car is then transferred outdoors, making it even hotter outside. The energy to power the air conditioners is likely to come from **fossil fuels** (coal, oil, natural gas). They're called fossil fuels because they come from the remains of plants and animals that lived millions of years ago and have been stored in the earth ever since. As they burn, they also add heat to the atmosphere because most fossil fuel power stations emit two-thirds of their heat as waste. In addition, they vent more CO_2 into the atmosphere, which adds to the greenhouse effect and further increases the temperature. As it gets hotter, more and more air conditioners are turned on, and they run longer (they also become less efficient in cooling). So if we try to combat global warming with air conditioners, we'll be creating a runaway positive feedback loop that won't work.

THINKING IN TERMS OF SYSTEMS SURE ALERTS YOU TO CONSIDER MORE UNINTENDED CONSEQUENCES THAN LINEAR THINKING DOES.

Chapter 5

Matter and Energy Cannot Be Destroyed

WHY ARE THERE SO MANY UNINTENDED CONSEQUENCES?

They are built into the laws governing natural systems. For example, the first law of thermodynamics (sometimes called "the law of conservation of matter and energy") states: **Matter and energy can neither be created nor destroyed; they can only be transformed**. The law sounds abstract, but it really says a lot more than meets the eye. The earth is a semi-closed system. Energy can enter the earth from space (most of it comes from the sun), and it can be radiated back to space. But matter stays on the earth and circulates internally within the earth's systems. **Matter cannot cease existing**, although it can be recycled. If you try to throw something away, it may leave your hand or your household but it always goes somewhere. So environmentalists are always saying, **"Everything has to go somewhere"**.

On television news recently a politician was urging that a toxic dump be "cleaned up". The reporter only asked where the money would come from to pay for cleaning up the dump. It never occurred to her to ask where the bad stuff would go. Because we can't create or destroy matter, we should always ask, **"Where did it come from? Where will it go? What will it do when it gets there?"**

IT SEEMS TO ME THAT TRYING TO SOLVE ONE PROBLEM CREATES ANOTHER.

And you've just derived another basic principle from the "first law". Environmentalists say it this way: **We can never do merely one thing**. Usually it doesn't stop with only one more thing but sets a whole chain of effects in motion. That leads to another principle from the "first law": **Everything is connected to everything else**.

YOU DON'T MEAN LITERALLY EVERYTHING?

I don't mean that you have to think of literally everything before taking an action; no one could do that. But when we act without thinking of unintended consequences, we may create bad things that nobody wants. For example, the way to my cousin's house in New Delhi is just adjacent to where a sprawling slum begins. The stench as we go to his house is unbearable, and never fails to elicit irritated comments. Everything changes as we enter the gate-barred middle class colony. His house is comfortable and pleasant, even if a little small. Like a typical upper middle class house, it is well lighted and ventilated and thankfully (as he says) at the other end from the slum. Like a typical upper middle class person, he leaves a lot of lights on all over the house. He is unconcerned about the wastage except to the extent that it affects his electricity bills. For political reasons, electricity is subsidized in India and the billing does not represent the true cost. Yet millions of people wasting electricity like this (even if they can afford the bills) means more of the familiar power shortage, means more demand for power stations from the politically powerful urban middle class, means more dams, means more displacement of people (as dams inundate huge areas of land and there is hardly any unoccupied land in India; Rajasthan, for instance, is the most thickly populated desert in the World) who often end up in city slums to reestablish their devastated lives.

And all the while we only wanted to have the comfort of not having to bother about switching lights and fans on and off. We never intended to add to the misery of the rural poor or to contribute to the growth of slums.

Another example from Nature shows how even scientifically we don't understand all these interconnections. Scientists in The Netherlands recently discovered that some bird eggs were not hatching because the eggshells broke in the nests. They traced this problem back to burning fossil fuels. The birds get the calcium for their eggshells from snail shells. The snails get the calcium for their shells by ingesting soil particles. Recently, however, the soils had become more acidic from acid rain. The rain had become acidic because of pollutants emitted when fossil fuels are burned. Acidic soil moisture dissolves calcium, thus depriving the snails of their calcium, which in turn deprives the birds of calcium and makes their eggs fragile. And again all the while humans only wanted to get energy from burning fossil fuel. They never intended to make acid rain or to make it impossible for birds to reproduce.

Chapter 6

Driving a Car Has Multiple Consequences

HOW DOES BURNING FOSSIL FUELS MAKE RAIN ACIDIC?

What happens when you burn a litre of petrol in your car? What does it do? Where does the petrol go?

THE PETROL MAKES THE CAR TRAVEL 10 TO 20 KILOMETERS AND THE PETROL JUST DISAPPEARS.

Does it? Let's take the principles from the first law of thermodynamics and apply them to the burning of a litre of petrol. Most of the chemical energy in the petrol is explosively transformed to heat; that's what gives the engine its power. The engine gets hot, but the radiator keeps it from getting too hot by dumping the heat into the atmosphere. The exhaust also gets very hot and dumps more heat into the atmosphere. Additional heat is generated by the friction of the moving parts and by the tyres gripping the roads. Notice that all of that heat comes from the petrol.

BUT THE HEAT ISN'T A PROBLEM. DOESN'T IT JUST BLOW AWAY IN THE WIND?

Sometimes the wind blows, sometimes not, but the heat from the burning petrol does raise the surface temperature of the planet. And it becomes a huge problem when lakhs of cars in a city like Delhi are burning lakhs of litres of petrol. The heat along the roads is tremendous. It's so hot that more and more cars have air conditioners. Ironically, the air conditioners take more energy, and generate more heat, just to cool the occupants of the cars. In a traffic jam, the heat buildup is almost unbearable. Most big cities, like Delhi or Los Angeles, are largely concrete and asphalt that absorb more solar heat than do green trees and lawns. So when the heat from the burning petrol is combined , most cities become "heat islands" that are 4 to 6 degrees centigrade hotter than the countryside.

THAT'S A HUGE UNINTENDED CONSEQUENCE OF PEOPLE MERELY TRYING TO GET SOMEWHERE.

That's only the beginning. Some of the burned petrol turns into new toxic chemicals that go out the exhaust. Air pollution in most big cities is so bad that people are harmed just by breathing. Another problem arises when some of the chemicals rise into the atmosphere, are transformed by the sun, mix with water vapor, and become smog. On most days, people in Delhi can hardly see the stars in the night sky because of smog. In fact a friend of mine had to give up astronomy as a hobby a few years ago. It was getting too inconvenient for him to always go to the outskirts of the city for his star gazing.

BUT THAT'S A SPECIAL PROBLEM IN DELHI, OR PERHAPS CALCUTTA TOO. THE REST OF THE COUNTRY ISN'T LIKE THAT.

Actually, quite a few cities, not only in this country but all over the globe, have severe air pollution problems from burning fossil fuels. But the damage isn't limited to big metropolitan areas - it reaches everywhere. As the chemical mixture emitted by cars, trucks, furnaces, factories, and electric power plants is blown by the wind and mixed with clouds high in the atmosphere, new chemicals are produced, many of which are acidic. These chemicals can travel hundreds, even thousands, of kilometers from their source. Eventually, these new and old chemicals fall back to earth, usually as some kind of precipitation. This precipitation, popularly called "acid rain", kills fish in lakes, injures trees in forests, erodes statues and stone buildings, and ruins the finish on cars. And it even prevents some birds from reproducing.

BUT WE ARE PASSING TOUGH LAWS AND DEVELOPING NEW TECHNOLOGIES TO STOP AIR POLLUTION.

We are, but that isn't enough. Tough laws create reactions that are hard to overcome. Remember, **we can never do merely one thing**. In the 'Inspector Raj' of India, tough environmental laws for industrialists often mean merely an extra Government inspector to bribe. Recently the Government made it mandatory for any new car sold in the four metros to be fitted with catalytic convertors - a device that cuts down heavily on pollution. But such cars require special lead free petrol. If one has to drive out of these cities, it is very difficult as lead free petrol is just not available elsewhere. Many people are reported to have gotten the pre-fitted catalytic convertor bypassed in their cars. Individualist thinking focuses narrowly on what is good for the self irrespective of consequences to society.

Another law requires all vehicles in Delhi to obtain Pollution-Under-Control certificates. Some time ago the period of validity of these certificates was reduced from 1 year to 3 months to make this check more effective. What people don't realize is that pollution is not just a measure of one parameter. All that these certificates certify is that idling CO (Carbon monoxide) emission levels are within norms. Nobody has a clue as to how these norms compare with those followed in the rest of the World. Besides, in the social climate prevailing in India today, laws such as these are at best a financial irritant. The PUC certificates are reportedly delivered at home for those who are willing to pay an extra price for this 'service'.

Despite our vigorous efforts to stop pollution, there is still a great deal of it, and we keep finding new pollution threats. In fact, despite having perhaps the toughest set of environmental laws in the entire third world, and having the largest technical manpower that could produce improved technologies for us, we seem to be losing the battle of enforcement.

Besides, don't forget that the way things are growing, we will be adding 600 vehicles a day in Delhi alone over the coming 10 years.

WELL, RECENTLY I READ ABOUT AN ELECTRIC CAR THAT DOESN'T POLLUTE. WHY CAN'T WE MAKE MORE OF THEM?

There's a big research and development effort going on now in the West to build better electric vehicles. There are laws, such as in California, that are spurring such innovations. The Californian law, for instance, requires that, by 1998, 2 percent of the vehicles sold by each manufacturer must have zero emissions; that requirement increases to 5 percent by 2001 and to 10 percent by 2003. So far, the only zero-emissions vehicles are electric. Governments and private companies are working hard to meet those deadlines, and I read recently that they're making good progress too. But even there the big car companies are resisting such efforts. They recently asked to be exempt from meeting the 2 percent requirement, arguing that the tougher standards were too difficult to meet and were not necessary, because they believed they could develop more efficient and cleaner petrol-burning cars and cleaner petrol as well.

In India, this is almost a non-option at the moment. There are no electric cars commercially available, and even if there were, their cost would be prohibitive and their market small. In fact the Indian market is so price sensitive that public transport authorities prefer to buy buses mounted on truck engines for a mere 8% reduction in price, even if this design mismatch means much higher pollution levels. If we can't even pay a little extra money for the benefits of relatively cleaner air, then there is slim chance of our adopting electric cars which are bound to be expensive at least in the immediate future.

Electric cars really aren't totally free of pollution either. Most of the first generation of electric cars will use lead acid batteries. Much additional lead will be mined and processed to make new batteries and lead will be reprocessed from exhausted batteries. Some of this lead, which is highly toxic, will be released into the environment no matter how carefully those actions are taken. We desperately need a breakthrough in battery technology. In addition, the electricity has to come from somewhere - usually a fossil fuel burning power plant. For instance, 60% of the electricity produced in India is generated in thermal power plants using coal. Those plants also pollute and put out a lot of waste heat, as I mentioned earlier; but the fuel is burned more completely there than in a car engine. It is also easier to control pollution at the plant than in lakhs of individual cars. Natural gas burns more cleanly and makes less CO_2 than coal or oil, so it's really the fuel of choice to fight air pollution.

Interestingly, in an advanced country like Holland, almost half the trips made by people are on bicycles.

I GUESS I'M LIKE OTHER PEOPLE. I LIKE THE INDEPENDENCE AND CONVENIENCE OF HAVING MY OWN CAR; I'D HATE TO HAVE TO CARPOOL OR TAKE PUBLIC TRANSPORTATION. MOST PEOPLE PROBABLY DON'T WANT TO KNOW ALL THE COMPLICATIONS THAT ARISE FROM DRIVING THEIR CARS.

Studies of the way people think about the environment suggest they don't. That's why we will be forced by regulation to adopt measures as in Singapore where limited car ownership rights are auctioned to the highest bidders, and there is a fine on cars entering the city centre if they are carrying less than 4 people.

Ironically, there's more to the story. Petrol comes from crude oil that was created and stored in the earth lakhs of years ago. Humans have already burned up close to half of it, just in the past 50 years. And at the rate we're going, it won't last another 50 years.

BUT WHAT WILL WE DO WHEN IT RUNS OUT?

Regrettably, we've grown so dependent on oil that running out will force wrenching changes in our lifestyles and economy. Most of our leaders are trying not to think about it, which only adds to the problem. It puzzles me why so many leaders of industry and government vigorously fight efforts to require us to use energy more efficiently. We're only hurting ourselves by refusing to conserve.

YES, I'VE SEEN THAT MYSELF. I'VE NOTICED THAT SOME PEOPLE IDLE THEIR CARS, WHILE PARKED, TO KEEP THE AIR CONDITIONER GOING IN THE SUMMER OR THE HEATER GOING IN THE WINTER OR JUST BECAUSE THEY ARE TOO LAZY TO START IT UP AGAIN WHEN THEY HAVE TO MOVE. IT NOT ONLY WASTES MONEY BUT ALSO USES UP ENERGY OUR GRANDCHILDREN WILL NEED TO LIVE.

And pollutes, too.

Chapter 7

The Greenhouse Effect and Global Warming

Earlier, I mentioned the greenhouse effect and global warming, but I didn't explain them very thoroughly. Another important reason for carefully conserving oil is that it forms carbon dioxide (CO_2) when it's burned. The amount of CO_2 in Earth's atmosphere rose from 270 parts per million (PPM) in pre-industrial times to 315 PPM in 1958 to 350 PPM in 1990. The U.S. Environmental Protection Agency, in 1985, predicted that if we follow our present trajectory of human activities, atmospheric CO_2 will more than double by 2025.

HOW SERIOUS IS THAT?

Actually, it's one of our most serious problems because it's likely to change the climate. The surface temperature of our planet really is very delicately balanced. If all the solar energy striking Earth were to remain here, the planet would be unlivable. Fortunately, most of the heat is radiated back to space. Our atmosphere intercepts some of the rays heading back to space and captures just enough heat to provide our normal surface temperature. We call that the **greenhouse effect**. A greenhouse lets heat rays enter, but it also traps some of them so they don't radiate out again. CO_2 is one of the most important greenhouse gases that intercepts rays headed back to space.

When you unlock your car after it's been sitting in a parking lot with the windows closed on a sunny day, it feels like an oven inside because it was acting like a greenhouse.

BUT WHEN I LEAVE A CAR WINDOW OPEN, SOME OF THE HEAT CAN ESCAPE. IS THAT WHAT HAPPENS IN OUR ATMOSPHERE?

Yes, it is, and the amount allowed to escape is crucial for life systems. Other planets have a greenhouse effect as well. Mars, for example, has almost no atmosphere and therefore has only a tiny greenhouse effect. It's hot in the day but very cold at night, because most of the heat was radiated back to space. So far as we know, there's no life there.

Venus is just the opposite, because its atmosphere is 97 percent carbon dioxide. Its greenhouse effect is so strong that it captures most of the heat from the sun. Scientists estimate that its surface temperature is somewhere between 400 and 500 degrees centigrade. That's hot enough to melt lead. We doubt there's any life there either.

The amounts of water vapor and CO_2 in the earth's atmosphere are the main regulators of the surface temperature on our planet. If they intercept too much of the solar radiation reflected from the earth's surface, we'll be too hot. If they intercept too little, we'll be too cold. We can do very little to control the amount of water vapor in the air, because it's mainly determined by the weather. But the amount of CO_2 is very much affected by what we do. There are other greenhouse gases too, but CO_2 is the main regulator over which we have some control.

Earlier, we talked about the danger in positive feedback loops. Today, there's a contemporary possibility that a positive feedback loop is getting underway in the carbon cycle. Many weeks of hot dry weather make forest tinder dry. Forest fires are a natural phenomenon, but its incidence can vary due to many factors, dry tinder and added heat forming a potent combination. Every tree that burns adds more CO_2 to the atmosphere. The loss of the photosynthetic action of the dead trees means that they can no longer take up carbon. Both actions increase the amount of CO_2 in the atmosphere, adding to the greenhouse effect, further warming the planet, and making it even more likely that forests will burn in the future. This in turn will add to the CO_2 and the problem will continually worsen.

So burning fossil fuels not only adds heat directly to the atmosphere and pollutes the air but also adds to the greenhouse effect. Pollution control equipment doesn't remove CO_2; it can only be curbed by limiting burning. So, as CO_2 builds up, the earth gets hotter and hotter. You often hear this referred to as **global warming**. The main thing we can do to avert further global warming is to use much less fossil energy.

IS GLOBAL WARMING RELATED TO LOSS OF THE OZONE LAYER? SOME PEOPLE CLAIM THAT HUMAN ACTIONS COULDN'T POSSIBLY HAVE CAUSED HUGE LOSSES OF OZONE. SOME EVEN BLAME IT ON VOLCANIC ERUPTIONS.

You raise two questions. First, those are separate but huge problems. Second, I can assure you that these people are wrong. The science on the loss of the ozone layer is very solid in attributing that loss to human actions. The evidence supporting that will make more sense if I explain some basic atmospheric science.

Ozone (O_3) is not a very stable element compared to **oxygen** (O_2) and is much more reactive. That means that it's more harmful to living tissue. It's important to distinguish ground level ozone from stratospheric ozone. Ground-level ozone is a serious pollutant and health risk in many large metropolitan areas and it's a major contributor to smog. It's created by burning fossil fuels, especially in motor vehicles. Under the latest U.S. Clean Air Act, for instance, they are trying to reduce ground-level ozone by requiring cleaner-burning vehicles as well as cleaner-burning fuels and by avoiding petrol evaporation at petrol pumps.

On the other hand, we **want** ozone in the upper stratosphere. A layer of ozone extending from 20 to 30 kilometers above the earth's surface intercepts ultraviolet B (UV-B) radiation from the sun. Without the ozone shield, increased UV-B radiation reaches the earth's surface, increasing the incidence of skin cancer and eye cataracts, weakening human immune systems, and injuring plants. Radiative damage is especially serious for tiny plants called *phytoplankton* that grow on the surface of the ocean, because those tiny plants are the foundation for the ocean food chain.

Now scientists are sure that **chlorofluorocarbons** (**CFCs**), invented by scientists in the 1930s, are destroying the ozone layer. At first, CFCs seemed like miracle chemicals because they were so stable and nontoxic. Engineers found many uses for them, for example, as propellants in spray cans, as fire suppressors in fire extinguishers, and as solvents with hundreds of uses in factories. They were an ideal refrigerant and were used everywhere in refrigeration and air conditioning units; chemical factories churned them out by the ton.

Alas, **we can never do merely one thing**. In the early 1970s, chemists at the University of California at Irvine discovered that when CFCs were released into the atmosphere, they rose gradually to the stratosphere, where solar radiation breaks apart those complex molecules. This chemical reaction releases chlorine atoms that are attracted to O_3 and break it down to O_2. As the chlorine atom reacts with O_3, the chlorine atom is freed up once more and can react with additional O_3 molecules - upward of 50,000 times. Chlorine seems to consume ozone molecules.

Even though UV-B breakdown of CFCs and the subsequent ozone devouring by chlorine atoms was demonstrated in the laboratory, the theory of ozone layer depletion was vigorously disputed, probably because so much was already invested in heavy use of CFCs. Extensive atmospheric monitoring got underway. Then, in the mid-1980s, scientists discovered a huge hole in the ozone layer over Antarctica; later, a smaller hole was discovered over the Arctic. Many scientists and political leaders took this as convincing evidence that preventive action must be taken, and a treaty was signed in Montreal in the late 1980s to phase out the use of CFCs. A later agreement in London accelerated the phaseout; all CFC uses are to be eliminated by the year 2000. Even so, there are millions of tons of CFCs still in use; much of it will escape, and the thinning of the ozone layer will continue for several decades.

Despite the widespread agreement among scientists and industrial leaders that CFCs must be phased out and more benign chemicals must be used in their place, many conservatives still claim that there is no truth to the O_3-CFC theory. Rush Limbaugh, for example, claims "Mount Pinatubo in the Philippines spewed forth more than a thousand times the amount of ozone depleting chemicals in one eruption than all the fluorocarbons manufactured by wicked, diabolical, and insensitive corporations in history... Conclusion: Mankind can't possibly equal the output of even one eruption from Pinatubo, much less a billion years' worth, so how can we destroy ozone?"

Scientists have been carefully monitoring the impact of Pinatubo's eruption and clearly disprove Limbaugh's assertion. "Cumulatively, Pinatubo's destructive effect on the ozone layer has been about 50 TIMES LESS than that of CFCs, rather than a thousand times greater, as Limbaugh claims. Thus, his estimate is off by a factor of 50,000.

On December 20, 1994, U.S. National Aeronautics and Space Administration (NASA) scientists held a press conference to release the findings of a new study from their Upper Atmosphere Research Satellite Program, which monitored the atmosphere above the Antarctic. The study's authors claimed to have definitively shown that the hole in the ozone layer is attributable to use of CFCs and not volcanic eruptions. Instruments in the satellite detected fluorine as well as chlorine in the stratosphere. "There is no natural source of fluorine" - thus it had to come from CFCs. In September 1995, 80 scientists from 26 countries concurred in a peer review of a report on ozone issued by the World Meteorological

Organization and the United Nations Environment Program. The report clearly attributes ozone loss to human-made chemicals, especially CFCs.

So, as you can see, loss of the ozone layer and global warming are huge but separate problems. Ground-level ozone and CFCs are both greenhouse gases, which may lead some people to confuse the two problems. There is widespread agreement that both gases should be diminished as quickly as possible.

Unfortunately, agreement on taking appropriate actions to avert global warming is more elusive than the agreement to phase out CFCs. Even though the theory that increasing levels of greenhouse gases will bring on global warming is taken very seriously by most atmospheric scientists, there are some scientists who doubt that global warming and climate change will actually happen. When faced with the dilemma of who to believe, we should ask, **"What is the cost of being wrong?"** Even if we cannot be absolutely certain that Earth systems will react the way the theory predicts, the consequences of climate change will be so dangerous and disruptive that we should not wait to find out if the theory is correct. The more ominous the consequences will be, the more cautious our policy should be. The most cautious action we can take is to cut back sharply on the use of fossil fuels.

BUT WOULDN'T IT BE EASIER TO TRY TO ADAPT AS WE GO ALONG RATHER THAN UPSET THE ECONOMY BY CUTTING WAY BACK ON BURNING FOSSIL FUELS?

Merely trying to adapt is just what I'm afraid of. That could be a big mistake, because we can't be sure that the effects of warming will come on gradually and be evenly spread.

Living things are very sensitive to climatic conditions. Humans and other animals can migrate away from areas that develop inhospitable climates, but plants can only die. Without food from plants, animals and insects also die. Plants especially cannot tolerate a climate that's too hot or too cold, too dry or too wet. The plants in a given area have adapted to the climatic conditions that have existed in their habitat for thousands of years.

Most climatologists expect the warming to be faster in the far north and south than near the equator. More importantly, as the heat trapped by CO_2 warms the oceans, it could change the direction of ocean currents. That, in turn, would change the direction of the jet stream in the upper atmosphere and change climate patterns. The weather isn't likely to just get warmer and settle down. The extra heat is more likely to drive the weather to extremes, creating floods, droughts, and violent storms. It may even oscillate. If we drive the climate further into chaos, it could change almost everything about the way we live.

THAT'S A PRETTY STRONG STATEMENT! ARE YOU REALLY SERIOUS?

I'm deadly serious. We're unintentionally conducting a gigantic global experiment to see how far we can disrupt global systems before they behave chaotically and destroy societies all over the globe. Do we want to continue our greedy consumption of energy and take the chance of destroying our own society as well as other societies around the world?

BUT YOU MAKE THIS THREAT SOUND AS BAD AS NUCLEAR WAR. COULD IT REALLY BE THAT BAD?

Exactly that comparison was made by a conference of atmospheric scientists from 42 nations who met in Toronto in June 1988 to discuss the changing atmosphere. They said that the threat of global warming was second only to nuclear war and urged nations to take precautions to limit emissions of greenhouse gases. It's truly incongruous that the world over, countries have been willing to spend trillions of dollars for armaments to avert a possible nuclear war but so far have refused to confront a threat that could be just as devastating and is more likely to happen.

WHY DO WE TAKE ONE THREAT SERIOUSLY BUT NOT THE OTHER?

I can only speculate, but I believe one reason is that we still haven't learned the basics of environmental thinking. Another reason is that most world leaders (who have a deep special interest in the economy) have convinced us that building up armaments stimulates the economy and that reducing consumption of fossil fuels will slow the economy. They tell us that growth in the world economy must come first and urge us not to accept policies that could slow it.

Following that same logic, however, I suggest that we put as much investment into harnessing solar power as we put into armaments. The boost to the economy would be similar to an armament buildup and would have the additional benefit of reducing the greenhouse effect. This lack of clarity in our thinking makes our society terribly vulnerable in the long run. Furthermore, our grandchildren will pay the cost of our failure to think clearly and to plan for the long term.

I SUPPOSE THE OIL AND COAL PRODUCERS WOULD FIGHT ANY EFFORT TO BUILD SOLAR POWER STATIONS THAT COULD STEAL SOME OF THEIR BUSINESS. BUT WHY SHOULD THE REST OF US LET THEM STOP US FROM CHOOSING WISE POLICIES?

You're right that such questions always get back to politics. I told you this story to illustrate how the innocent act of burning a litre of petrol has a multitude of effects, which, when multiplied by billions of people doing it, can have the cumulative effect of drastically changing the way the world works. The essence, then, of environmental thinking is to constantly ask, **"And then what?"** The most important thought to remember from this conversation is to **keep asking, "And then what?"** It is doubly

Chapter 8

Scattering Matter and Energy

WHEN WE DISCUSSED THE FIRST LAW OF THERMODYNAMICS, YOU SAID THAT EVERYTHING HAS TO GO SOMEWHERE. BUT WHEN THINGS BURN OR ROT, WE ONLY SEE REMNANTS. WHAT HAPPENS TO THE OBJECT? WHERE DOES IT GO?

That's explained further by the second law of thermodynamics, also called **the law of entropy. Entropy is a process by which energy gets transformed and scattered; it always happens when work is done.** When petrol is burned, the energy goes through the entropic process. It's scattered in many places and can no longer do any work. Unlike matter, energy can't be recycled; you could say that it always "runs downhill". The entropy law was written to apply only to energy, but something like it also applies to matter. Even something as solid as steel will rust away and become scattered and useless. All products eventually become scattered. When living things die they decay and become scattered. Matter can be reassembled into something useful by applying more energy to it; that's what we do when we recycle glass, metal, plastic, and paper. But the energy expended in recycling becomes entropic. All energy is eventually degraded to random vibration that can't be recovered for work.

SO IF EVERYTHING IS WINDING DOWN, GOING THROUGH ENTROPY, AS YOU CALL IT, WHAT KEEPS THE WORLD GOING?

Well, if it weren't for the energy from the sun, everything on Earth would wind down. As I said earlier, plants use chlorophyll to take energy from the sun, carbon from the CO_2 in the air, and water containing nutrients from their roots to make new structures in a process called **photosynthesis**. The new structures (leaves, stems, seeds, fruit) contain bound-up energy. So photosynthesis is the main force countering entropy that keeps life systems from winding down. Humans, as well as all other creatures, live off this creative ability of plants. We get nearly all of the energy we use from the sun. Even coal and oil were made aeons ago by plants using solar energy.

YOU SAID WE GET NEARLY ALL OUR ENERGY FROM THE SUN. WHERE DOES THE REST COME FROM?

We've found energy three other ways. In a few places, heat from the interior of the earth is close enough to the surface to allow people to tap into it and make it do work. That's called **geothermal energy**.

BUT HOW CAN THEY DO THAT?

Geologists know that in places where there are geysers or volcanic craters the molten mantle is fairly close to the surface. A hole is drilled to hot rock. Water that is pumped down the hole turns to steam as it contacts the hot rock. The steam is piped to a turbine that powers an electric generator. Another non solar source is **tidal energy**. Tides are made by the gravitational pull on our planet by other planets and the sun; but the strongest pull comes from the moon. In some places on Earth - the Bay of Fundy on the Canadian east coast, for example - the tides are high enough and the water moves swiftly enough that a dam can channel the moving water to turn electric turbines. But the total amount of energy from these two sources is quite small.

A third source is **nuclear power**, which gets its energy from the fission of unstable uranium. Fissionable materials have quite a lot of usable power but they're risky to use and create dangerous wastes that we don't yet know how to store safely. The problems have been so great that, even after 30 years of nuclear buildup, the world gets only about 5 percent of its energy from nuclear power. As you know, while there's vigorous debate about whether or not we should develop more nuclear energy, a developed country like Sweden, where half the electricity is generated from nuclear power presently, has already decided to phase it out completely over time. It's a complicated question about how much risk we should take and whether the benefits outweigh the drawbacks.

WHAT ABOUT **HYDROELECTRICITY**?

It too comes from the sun, which evaporates moisture from oceans, lakes, and plants. The moisture forms into clouds which are carried by the wind (whose energy also comes from the sun) to high ground or mountaintops. The rainwater is collected into rivers and lakes by gravity and the weight of the water as it flows downhill can spin a turbine to make electricity - that way we can recover part of the solar energy that made the water evaporate in the first place. As long as the sun shines, solar energy is inexhaustible and non polluting.

Actually, water is a potential source of energy in another way, because it's made up of two hydrogen molecules and one oxygen molecule. If the hydrogen and oxygen are separated, the resulting hydrogen gas could then be burned (that is, recombined with oxygen), producing intense heat that could be harnessed to do work. Burning hydrogen doesn't produce pollution either, only water. Water is also one of the most plentiful substances on the planet. Unfortunately, the only way we know how to separate hydrogen from oxygen is to pass an electric current through water in a process called electrolysis. We haven't yet figured out how to do that economically, because it takes more energy for electrolysis than can be recovered when hydrogen is burned. Scientists and engineers are working on plans to generate electricity with large arrays of **photovoltaic cells** set up in deserts and hope to use that electricity to separate out hydrogen. The hydrogen can be bottled or sent in pipelines to consumers. It's a way to move energy from a place where there's a surplus to a place where humans really need it - to cool homes or to power cars, for example.

WHAT ARE PHOTOVOLTAIC CELLS?

They're made from layers of silicon, which is one of our most abundant minerals. Light energy stimulates a small electric current between the silicon layers. Many hand-held calculators, which require only tiny amounts of electricity, are now powered by photovoltaic cells. Because the output from each cell is comparatively weak, it takes large arrays of cells to get enough electric current to make the electrolysis process work. Even though this is a relatively expensive way to make electricity, using the excess solar energy in deserts could make it worthwhile.

THAT WOULD BE A NICE CLEAN WAY TO TURN SUNLIGHT AND WATER INTO USEFUL ENERGY. ARE THERE ANY OTHER POSSIBILITIES?

Scientists are working on another way to get energy from water, called **fusion power**. Deuterium and tritium are two isotopes of hydrogen present in small proportions in sea water. Since there's so much sea water, there would be plentiful amounts, and they can be separated out quite cheaply. Scientists have named this fusion energy because heat is released when the isotopes are brought together and fused. The fusion reaction is instantaneous and creates heat as intense as the surface of the sun. It's the same energy that is released in hydrogen bombs.

Unfortunately, it requires temperatures ten times that of the sun to initiate the reaction. Scientists have been trying for several decades, without success, to initiate, sustain, and control the reaction so the energy can be used to do work for humans. It's anybody's guess as to whether or not fusion will ever become an energy source for everyday work.

GETTING ENOUGH ENERGY IS REALLY COMPLICATED, BUT I CAN SEE THAT MANAGING ENERGY BETTER IS A KEY TO SOLVING ENVIRONMENTAL PROBLEMS.

Actually, the fastest and cheapest way to get more energy is to use more efficiently the energy we now have to do the things people want - lighting, heating, cooling, transporting people and freight, pumping water, turning motors, and so forth. We can achieve greater efficiency by using more efficient new technologies and by changing the way we do things. I installed a new solar water heater in my home, for example. It cost us about Rs.15000 and saves us about Rs.900 a month on winter power bills. Assuming that I use hot water for only 4 winter months in a year (Nov - Feb), we save Rs.3600 a year, which works out to 24% return on initial investment. Not only that, electricity cuts don't affect me any more if I want a warm bath any time of the day or night. A number of studies suggest that Americans, for instance, could have the same, or an even better, quality of life and use less than half the energy they now use. Japan and most western European countries use only half the energy per capita that they do.

You may be surprised to discover that despite being privately owned, electric power companies in the U.S. are using special incentives to encourage people to use <u>less</u> electricity. Scientists have convinced them that getting people to conserve will save power companies the huge cost of building new power plants. The power companies actually come out ahead financially by giving away new energy-efficient fluorescent light bulbs.

Finding so many sources of energy is a great human achievement that has done a lot to improve our quality of life. Our job now is to refine our uses so that we don't exhaust our sources and don't interfere with the planet's biogeochemical systems. Solar energy is the safest underdeveloped source with the greatest potential for lasting indefinitely. There are lots of clever ways to put solar energy to work, and we should direct more effort to doing that. At the same time, we should avoid using more and more fossil energy (oil, gas, and coal), because it's in limited supply and creates pollution and global warming problems.

EXPLAIN WHAT A **BIOGEOCHEMICAL SYSTEM** IS.

"Biogeochemical" is a shorthand word scientists use for three intersecting systems: biological systems, geological systems, and chemical systems. They're all crucial for sustaining the flourishing life found on Earth. These systems are intricately balanced and interconnected. Interference with any one of them has repercussions in the others. I won't go into detail because they're very complicated. Since these three systems are so crucial for life, I often refer to them collectively as **"life systems"**.

Chapter 9

Dynamic Natural Systems Stabilized by Diversity

I'M REALLY STRUCK BY HOW FINELY BALANCED AND EASILY DISTURBED LIFE SYSTEMS ARE. WHAT KEEPS THEM OPERATING SMOOTHLY MOST OF THE TIME?

Environmental scientists don't understand all the dynamics, but one key to the stability of natural systems is the diversity of plants, insects, and animals that live in them. If nature is left undisturbed by humans, it naturally evolves ecosystems with incredible diversity, leading to many interconnected negative feedback loops that suppress disruptive changes. Recent research shows that diverse plant ecosystems are not only more resilient and robust, they're also more productive and consume more CO_2.

Tropical rainforests are generally believed to have the most diversity on the planet. A tropical forest in Panama has more kinds of plants than live in all of Europe. A small wildlife reserve near the equator has over 500 species of birds and 800 types of butterflies. Almost half the plants there haven't been named. About 15 percent of all the species of birds in the world nest in the rainforests of Indonesia.

ISN'T THERE TERRIFIC COMPETITION IN ECOSYSTEMS THAT HAVE SUCH DIVERSITY?

By trial and error each species finds a **niche** in which it can live. We could call it their home. It is quite common for a species to develop **a symbiotic relationship** with one or more other species. When species help each other out, it's called **symbiosis**.

GIVE ME AN EXAMPLE OF SYMBIOSIS.

A common example is the symbiotic relationship between birds and cattle. You can often see birds feeding upon insects off the backs of indifferent cattle. The cattle provide a living environment to these insects. They need to get rid of them and the birds need them as food. Another good example is the relationship between bees and flowers. Bees are attracted to blossoms where they can get nectar to make honey. As they fly from flower to flower, they pollinate the blossoms so the plants can bear fruit. Neither bees nor flowers know that they are helping the other, but both get benefits.

This second relationship is so beneficial that humans have figured out a way to become part of the symbiosis. Beekeepers in the U.S. put their hives on large trucks, which they move from south to north, following the fruit orchard blossoming season. A bee truck parks at a blossoming orchard for a few days and the orchard owner pays the beekeeper a fee for providing swarms of pollinating bees. After a few days, the truck is moved at night to the next blossoming orchard while the bees are sleeping in their hives. The bees soon fill the hives with honey that the beekeeper harvests. Sale of the honey, plus the money collected from orchard owners, provide a good living for the beekeeper. The orchard owners are happy for their pollinated trees which then become laden with fruit. Presumably, the bees also thrive.

THAT'S CERTAINLY CLEVER. BUT HOW DOES DIVERSITY MAKE ECOSYSTEMS MORE STABLE?

I'll tell you a couple of stories. About ten years ago, I was hiking in a forest in western New York that hadn't been disturbed by humans for at least fifty years. I discovered something about stability through diversity when I noticed that all the birch trees had been killed by an insect called the birch tree borer. The dead trees were decaying, becoming food for insects and for new plant growth. Even though all the birch trees had died, many other tree species were thriving as were many insects and animals. The forest ecosystem was still flourishing.

On the other hand, when humans manage an ecosystem to maximize output, they typically create a **monoculture** - just one crop with as few weeds and insects as possible. In the short run, a simplified ecosystem gives humans rich returns of that one crop. But it's also very vulnerable. For example, U.S. farmers typically plant millions of acres of corn. By the early 1970s,70 percent of the hybrid seed corn came from only six inbred lines. Then an epidemic of leaf fungus swept through the cornfields, turning the corn leaves yellow and injuring the crop severely. American corn production dropped by 15 percent that year (by one-half in parts of the south), and many farmers lost their entire crop. After that, they learned to limit the damage by using seeds that contained a blight-resistant germplasm derived from wild corn found in Mexico. But of course corn is still vulnerable to other diseases and insects.

In India, paddy (rice) has always been one of the most important crops. Traditionally, thousands of varieties have been in use, sustained predominantly through the mechanism of culture, religious traditions and social practices. Then with Green Revolution techniques introduced in the sixties, efficiency and productivity based on science became more important than maintaining old traditions. As a result, old local varieties came to be rapidly replaced with new high-yield varieties with strong Government support. So much so that many varieties, developed over thousands of years of cultivation by humans, have already been lost forever. Many more are being lost every year even now owing to lack of cultivation, and only a few dozen varieties are in use in any appreciable quantities. And with this the danger of what happened with cornfields in the U.S. in the seventies increases.

Because any single crop is vulnerable to unexpected loss, farmers should plant diverse crops and rotate them on their fields. The researchers studying productivity from biodiversity have found that polycultures of about ten species produced the greatest bioproduct.

WHAT ABOUT CHEMICALS? AREN'T THEY MORE EFFECTIVE FOR KILLING INSECTS, WEEDS, AND AGAINST DISEASES?

Many farmers believe they're more effective and they're widely used. But now we're finding that the long-run use of chemicals has many unintended bad consequences. Insecticides kill all kinds of insects, not just those that injure crops. For example, the Ministry of Agriculture in Indonesia encouraged farmers to use more insecticides on rice in the early 1980s. But rice production fell rather than rose, because the brown planthopper became immune to the insecticides and increased in numbers while most of its natural enemies were killed off. In 1985, President Suharto banned 56 of 57 insecticides used on rice, and rice production gradually rose to former levels.

Good soil is alive with bacteria that perform many beneficial functions for plants. But soil laced with insecticides has many fewer bacteria. It's much less productive and erodes more easily. Also, birds that eat insects that have eaten chemicals get poisoned and die. Eventually, a few individual insects evolve an immunity to the insecticide and pass this immunity on to their offspring, thus their numbers begin to flourish again. But by then the reduced bird population is no longer able to keep them in check. New pesticides are then used to try to control the immune insects, so more and more kinds of chemicals are inflicted on the earth. Pesticides also may make farm workers sick, or maim or kill them. And these poisons often end up in the food in your vegetable markets.

BUT WHY DOESN'T THE GOVERNMENT STOP THAT?

Pesticides were used for decades before some scientists began looking for them in food. The amount of toxins found in food reaching markets was small, and the priority in India was on providing for an ever-growing population. A society that was not even self-sufficient in food grain production at the time of independence was not interested in complicating the issue by also looking at the quality of food produced. It is more damaging for any Government of the day to have starvation deaths in the country than to have people dying due to diseases caused by toxins in their food . The latter is not as obvious, and moreover, is debatable. Public officials, scientists and lobbyists in even the developed countries are still arguing over how much we can eat before toxins become dangerous.

An interesting sidelight here is the case of many big farmers like those in Punjab. They grow for the market with techniques as advanced as any in the developed world, but still keep aside a portion of their fields to grow food for their own consumption in a traditional manner. Without understanding the science behind it, they still intuitively believe that food produced in the modern way is somehow harmful, and so they reject for themselves what they produce for others to eat.

The whole story about chemical use is much more complicated than my simple examples. But my point is that **we can never do merely one thing**. When we try to manage complicated ecosystems to maximize their output for humans, we generally fail to ask, **"And then what?"** So we create unwanted effects that reverberate throughout the system.

Another unwanted effect is the **extinction of species**. The growing human population, using all the power science and technology has given us, poses another threat to biodiversity by stealing and degrading the natural habitat of other creatures. Humans now take about 40 percent of the planet's terrestrial bioproduct for their exclusive use. When we double our population, will we take 80 percent of it? And because of our relentless appropriation of biodiverse habitats for our exclusive use, species are becoming extinct at an alarming rate. There are hundreds of species on the Indian government's endangered species list. No one knows for sure how many species are being extinguished, because so many aren't even identified and named. But estimates of the numbers lost vary from many thousand to 15 million worldwide. There's no doubt that life systems on our planet are losing a priceless and irreplaceable genetic heritage.

Humans don't intend to extinguish species. I know of only one instance in which humans deliberately set out to eliminate a species: the smallpox virus. We thoughtlessly eliminate species because we focus so narrowly on satisfying our desires for food and shelter, fun, money, and power. For instance, in the U.S. the Endangered Species Act is now under attack for those motives. This is not only shortsighted but also morally wrong. The environmental philosopher Holmes Rolston III has written: "Every extinction is a kind of super-killing. It kills forms (species) beyond individuals. It kills `essences' beyond `existences', the `soul' as well as the `body', it kills collectively, not just distributively.... To kill a species is to shut down a unique story".

BUT DON'T WE HAVE TO USE MONOCULTURES AND CHEMICALS TO MAXIMIZE FOOD OUTPUT SO WE CAN FEED ALL THE HUMANS NOW LIVING AND ALL THE NEW ONES SOON TO BE BORN?

If you believe that humans should come first, please keep in mind that **a planet without suitable habitat for non-human creatures is not a suitable habitat for**

humans. Whatever we do to the rest of the world, we do to ourselves. Those who believe that humans should come first must recognize in their own self-interest that **our most core value should be the continued viability of life systems**. What's more, this is a deep moral issue. Can you say it's ethical to always give priority to humans? What about the welfare of all the other creatures who live with us on the planet?

BUT OUR CULTURE TEACHES US THAT HUMANS SHOULD COME FIRST.

Not at all. In fact, the Indian culture has always emphasized an attitude of caring towards all life forms regardless of their utility to man. What you seem to be referring to is the dominant culture today that has largely been shaped by aping the West.

Being human, it's very tempting to say that humans should come first. But that policy can't hold up over time. What kind of Earth will we have if humans try to take it all for themselves? Do you want your children to live in a world of 50 billion people, in a mainly artificial environment, without trees, lakes, birds, insects, and wildlife and with scarcely enough food and water to exist? Surely living should be something more than merely not dying. I'll go further; **if the human birth rate threatens the viability of life systems,** as is now happening, **it is morally wrong to encourage people to have more children.**

BUT THAT FLIES IN THE FACE OF A CHERISHED BELIEF IN OUR CIVILIZATION THAT CHILDREN ARE TREASURES. PEOPLE DON'T LIKE ANYBODY TO TELL THEM HOW MANY CHILDREN TO HAVE. MOST EDUCATED URBAN INDIANS PROBABLY AGREE THAT TWO CHILDREN ARE ENOUGH. WON'T THAT TAKE CARE OF THE PROBLEM?

It would help if most families accepted the moral obligation to stop with two children. But as you know, many do not. First, educated urban Indians make up an extremely slender minority. Even amongst them, there are many over whom the hold of tradition, or sometimes parents, remains strong. There are others who do not practice what they preach. They think that the poor having too many children is the problem, but not them. Coercive methods used in the past have made this such a touchy issue today that no political party is willing to touch it with a barge pole.

The threat to life systems from human population growth is so great that we must speak up. I deeply believe that we must accept the moral obligation to have fewer children and encourage others to do the same. To make my point even more strongly I should explain **carrying capacity**, which is another central concept of environmental thinking.

HOLD IT! I'M HAVING A HARD TIME DIGESTING EVERYTHING YOU'VE SAID SO FAR. BUT I DO FIND IT CHALLENGING AND INTERESTING. CAN WE COME BACK TO THIS TOMORROW MORNING?

Sure! Let's make it about nine in the morning.

Chapter 10

Limits to the Earth's Capacity to Support People

Morning, Madan. How do you react to our discussion so far, now that you have had time to reflect on it?

IT SURE GOT ME THINKING. IT WAS ESPECIALLY HARD FOR ME TO ACCEPT THE IDEA THAT HUMANS AREN'T MORE IMPORTANT THAN OTHER CREATURES. I THOUGHT ABOUT THAT A LOT LAST EVENING, ESPECIALLY YOUR POINT THAT IF WE DON'T PRESERVE NATURAL SYSTEMS WE WON'T HAVE MUCH OF A LIFE, PERHAPS NO LIFE AT ALL. I KEPT WONDERING HOW THE POLICIES YOU RECOMMEND COULD EVER BE PUT INTO PRACTICE. YOU AND I KNOW THAT MOST PEOPLE IN THIS COUNTRY THINK OF THEMSELVES AND THEIR OWN PLEASURE FIRST. HOW CAN WE ARGUE WITH PEOPLE LIKE THAT?

People who think only of themselves just aren't aware of how dependent they are on life systems and social systems. You might ask them if they could live without a viable ecosystem. Of course, once they think about it, they'll realize that their life is absolutely dependent on a properly functioning ecosystem that provides clean air and water, nutritious food, a gentle climate, and beautiful surroundings. Then you might ask them if they need a well-functioning society to lead a good life. Again, it's obvious that they can't be happy without it. Both of these systems must work well before humans have any hope of living a decent life. Since they also recognize that society can't function well without a well-functioning ecosystem, you can point out to them that they've worked out a simple value priority. **First priority must go to the ecosystem, second to the society; only when those two systems are well cared for, is it logical to emphasize personal desires.** Those priorities apply to making public policy as well as to making individual decisions. Such other values as making money, emphasizing economic growth, and having more children must take a back seat to preserving the viability of our biocommunity and our social community.

OK, THAT HELPS. NOW WHAT WAS IT YOU WERE BEGINNING TO TALK ABOUT YESTERDAY? WAS IT RELATED TO THIS TOPIC?

Yes. Ecologists speak of ecosystems as having a **carrying capacity**. That's the total bioproduct that the producers (plants) in that system can produce. An ecosystem with good soil, the right amount of rain, and the right temperature will produce much more bioproduct than a system that's deficient in one or more of those factors. Carrying capacity is very dependent on weather and climate.

In addition to thinking about the carrying capacity of small ecosystems, it's important to be concerned about the carrying capacity of regions, nations, and even the whole planet.

The number of any given species an ecosystem can support depends on how that species eats, and also on how all the other species in the system get their food. Humans are just one of those species. But we've been able to reproduce so quickly, especially over the last two hundred years, because we've taken over so many ecosystems for our own use and have manipulated them to maximize the output of human food. Scientists have also been able to develop plants and animals that produce more food for humans. And farmers maximize yields by spreading chemical fertilizer and using pesticides to keep down pests. They also use powerful machines that reduce labor.

The fertilizers, the pesticides, and the power to make and run the machines are a kind of subsidy to our food supply, with the energy coming mostly from fossil fuels. Then, the raw material for most fertilizers is crude oil. Experts estimate that in North America it takes about 5 litres of petrol to grow a bushel of corn or half a kilo of mutton. They also estimate that it takes ten calories of fossil fuel to grow, process, and deliver one calorie of food to your dinner table. How long do you think that can go on? Remember, we've used up more than half of the world's oil just in the past fifty years.

Also, you mustn't fail to consider that India's food grain production that has shown a remarkable rise has now stabilized and may not increase much while population grows relentlessly.

IT SURE LOOKS LIKE WE'RE LIVING ON BORROWED TIME.

Exactly! Since that subsidy of our food supply comes from fossil fuels, William Catton has called it **"ghost acreage"** and **"phantom carrying capacity"**. In other words, when fossil fuels are exhausted, the subsidy will disappear like a phantom and our food supply will decline quickly. Americans, for instance, are currently living off four parts phantom carrying capacity for every one part real carrying capacity.

Also, biologists who study the behavior of organisms in relation to their food supply have observed that when a species has plenty of food, it reproduces so quickly that it **overshoots** its food supply. When births exceed deaths, a population is growing. All population growth is exponential; that means it will double in some time period. How swiftly it doubles depends on the rate of growth. Biologists use the phrase **doubling time** to indicate the speed with which the population of a species doubles in size.

HOW DO THEY ESTIMATE THE DOUBLING TIME?

An easy rule of thumb is called **the law of sevens**, and it works this way. **Divide the percentage of growth per year into 70.** If a population grows 1 percent a year, it will double in 70 years. At 2 percent a year, it will double in 35 years. At 7 percent, it will double in 10 years. At 10 percent a year, it will double in 7 years. Remember the riddle of the lily in the pond. If the base is already high, the doublings really become huge.

Most species grow much faster than humans. When scientists put some kinds of bacteria into a petri dish with a good supply of food, their numbers will double in 15 minutes. If that went on for one day, even for something as small as bacteria, its volume would be as large as the planet. Of course that doesn't happen, because the bacteria soon consume their food supply, go into overshoot, and die back. The exhaustion of the food supply, and subsequent dieback, of a species may also injure the ecosystem so much that it reduces its carrying capacity.

CAN YOU GIVE ME AN EXAMPLE?

Sure. Australia was so far from other continents that mammals never evolved there.

I'VE SEEN LOTS OF PICTURES OF KANGAROOS IN AUSTRALIA. AREN'T THEY ANIMALS?

They're animals but they are not mammals; kangaroos and their relatives are marsupials. Unlike mammals that carry their offspring to full development in the mother's womb, marsupial young are born very small and live in the mother's pouch for many months until they can get along on their own. But let's get back to my example.

When the English conquered and colonized Australia, they introduced many mammals such as sheep and cattle. They also introduced rabbits so "gentlemen could have sport". But the rabbits had few natural predators in the Australian ecosystem, so they reproduced at epidemic rates. There were soon so many of them that they stripped the countryside of nearly all its vegetation as they went into overshoot. Of course, without plants to eat, other animals such as sheep, cattle, and marsupials died along with the rabbits. In fact, the rabbits died in such droves that they were burned in huge bonfires. The plant communities of the Australian ecosystem were devastated and took many years to recover. As they did, the rabbits once more began to reproduce epidemically and went through the whole overshoot and dieback sequence again.

As you can imagine, the humans got pretty desperate, so scientists introduced the myxoma virus, spread by fleas, that targeted only rabbits. That virus killed off 99 percent of the rabbits in the colonies it infected, and those ecosystems could recover. But to the humans' dismay, a few of the rabbits became resistant to the virus, and that

strain began once more to become a serious pest. Australian scientists are now searching for new fleas from Spain that can tolerate an arid climate so that they can transmit the virus more effectively. They're also genetically altering a virus so it will make rabbits sterile; that's now being tested. The English immigrants who introduced the rabbits didn't know that we can never do merely one thing and didn't ask, "And then what?"

But let's apply carrying capacity ideas to today's problems. The world is now experiencing a population explosion of humans. If it weren't for phantom carrying capacity, our species would already be in overshoot. So we should be asking ourselves, "Now what?" When we lose the fossil fuel; subsidy, we can be sure that we'll steal more and more productivity from other species, and that will drive many of them to extinction as we humans also head for dieback. But unlike the rabbits in Australia, we can foresee going into overshoot and dieback. **That foresight places a moral obligation on us to limit human reproduction and to moderate our consumption.** Usually, the poor produce more children and the rich consume more per capita. But in this simple moral obligation, they both see only the other's responsibility.

At the Global Environmental Summit in Rio de Janeiro in 1992, there was so much opposition to limiting human population that the topic couldn't even be brought up for discussion. Slogans like "Two cars for you, one child for us!" aptly described the 'Passing-the-buck syndrome' that prevailed. It's mind-boggling to me that humans can be so blindly stubborn about a matter of life and death - not only for us but for millions of other species.

Chapter 11

The Tragedy of the Commons

DO YOU REALLY THINK HUMAN SELFISHNESS AND SHORT-SIGHTEDNESS WILL SEND US TO OUR DOOM?

Van Rensselaer Potter called our unwillingness to use foresight and to take precautionary steps "evolution's fatal flaw". This flaw will send us to our doom if we continue to indulge the short-term desires of humans now living. We're also blindly selfish in another way that Garrett Hardin called **the tragedy of the commons**. He invited us to picture a village with a grassy common area at its center. Each villager has a right to graze one or more cows there. It's in the private interest of each villager to graze as many cows as possible on this common land, because the cost is shared by all the villagers, while he alone gets the milk from his cows. As each villager acts in his private interest, there are soon so many cows that the grass on the commons is destroyed. And therein lies the tragedy. The only workable solution to the short-sighted greed of each individual is to set up some form of governance to regulate the use of the commons. Our current challenge is to protect our global commons by developing a viable form of governance, even if we can't yet achieve a world government, that can overcome the fatal flaw in our biological evolution and avoid tragedy for the planet.

I COULD UNDERSTAND THIS TRAGEDY OF THE GLOBAL COMMONS BETTER IF YOU GAVE ME AN EXAMPLE.

Actually, examples of the tragedy of the commons are happening at many places now. For centuries humans could go to sea and catch all the fish they wanted without worrying about depleting the stock. But in the past few decades, large corporations have used special trawlers that tow huge driftnets, some of them more than 20 kilometers long, that scoop up everything in their paths. Fish are caught so efficiently that the breeding stock is exhausted, causing fish populations to drop dramatically.

Since there's no world government, some countries have taken action on their own to try to avert the impending tragedy. Many countries have already unilaterally extended their governance over the sea to 200 miles from their shores and are policing fishing in the waters under their claimed jurisdiction. A Law of the Sea Treaty, sponsored by the United Nations, recently went into effect under which all signing nations agreed to recognize the 200-mile limit. Most seafaring nations have now ratified the treaty. That's an example of the kind of governance that's possible without a government.

In 1991, Canada completely closed its Grand Banks cod fishery off the coast of Newfoundland, and threw 30,000 people out of work, in the hope that the absence of all fishing might enable the cod fishery to recover. Of course, that didn't prevent fishing beyond Canada's 200-mile limit. Trawlers from European Union countries, especially Spain and Portugal, continued to deplete the outer cod banks. In March 1995, Canada impounded a

Spanish trawler fishing beyond the 200-mile limit to prove that it used illegal small-mesh nets to scoop up small fish that are the main food for cod. This incident led to a new agreement between Canada and the European Union and to a discussion in the United Nations toward a new global fisheries convention. This story illustrates how difficult it is to avert the tragedy of the commons without an effective government. So far, fish breeding stocks haven't recovered and the Grand Banks fishery is still closed.

Fishing nations are now trying to outlaw the huge driftnet fishing trawlers that destroyed fish stocks. The U.N. General Assembly has twice passed resolutions urging nations to outlaw them, but some nations with large fleets are still resisting. By the time all fishing nations agree to outlaw them, it may be too late. There's no assurance that fish stocks will rebound when the trawlers are out of business.

As you may know, the action to protect whales was even stronger. Several years ago, by international agreement, all whale hunting was forbidden until there was clear evidence that whale stocks were recovering. Japan and Norway have recently resumed harvesting certain kinds of whales whose stocks, they claim, have recovered. But the other nations dispute that claim and may apply sanctions. International whale protection has been successful enough that the United States removed California gray whales from the endangered species list in the summer of 1994.

And here's yet another example of the tragedy of the commons. Before environmental awareness, air and water were treated as a common into which factories, businesses, and households could dump their wastes. As a result, air and water became so polluted that governments all around the world passed strict regulations over what and how much could be emitted into the air and water. **The more crowded we become, the stricter we must be in regulating what people can do with the commons.**

BUT PEOPLE HATE REGULATIONS; THEY'LL FIGHT THEM AND LITTLE WILL CHANGE.

Yes, I agree that people hate regulations. But we'd be lost without them, and they're often quite successful. For example, transportation would be a mess if we didn't have traffic rules. And do you remember, back in the 1970s in America, how people fought the regulation that car mileage had to be increased and exhausts cleaned up? The automobile manufacturers claimed that cleaner exhausts would require poorer mileage - that a fuel-efficient, clean-burning car couldn't be built. They said it couldn't be done but we did it. The 1993 models, on average, emitted less than 10 percent of the pollution of the models made in the early 1970s, while getting twice the mileage. Despite that, cars still emit lots of carbon dioxide as well as conventional pollutants, and with so many more cars taking to the road, there's still heavy pollution.

OK, IN THAT CASE THE REGULATIONS WERE SUCCESSFUL. BUT I STILL SAY PEOPLE WILL FIGHT REGULATION AND CHANGE.

You'd be right most of the time. But another American example will throw light on the issue from another angle. Way back in the 1950s, 1960s, and 1970s when nearly everybody took smoking for granted, I remember people being thought weird if they asked others not to smoke at a meeting or party. Anybody would doubt if people would ever change. Yet twenty years later, there's been almost a complete turnaround in North America. Most Americans readily accept regulations that forbid smoking in airplanes, buses, trains, and even restaurants, office buildings, and factories.

THAT'S A POWERFUL EXAMPLE OF UNEXPECTED CHANGE. MAYBE HUMANS CAN ACT TO SAVE THE EARTH.

Chapter 12

The Earth: Both Source and Sink

When humans want to make something, they must go to the **source** - the earth. They then process the resources they take from the earth into a product that's sold and used for a while. Eventually it's discarded, so the earth serves as a **sink**. Because the earth is both source and sink, we must be especially careful how we do things. Wastes are produced when the resource is extracted from the earth, when it's processed into a product, and when the product is used (remember the wastes produced when cars burn petrol). All of those wastes have to go into a sink. This flow of materials from sources to sinks is the **throughput** we talked about earlier. When our leaders call for faster economic growth, they're calling for faster throughput.

Because Earth is a finite planet, there are limits to both sources and sinks. There's only so much usable land, so much available oil, so much minable coal, so much recoverable iron ore, etc. The swifter the throughput, the faster the **stocks** of resources are depleted. Swift economic growth speeds us toward resource scarcity. But our leaders don't tell us that. Once a resource has been pushed through the throughput cycle, most of it becomes useless, but it still has to go somewhere - into a sink.

BUT CAN'T WE RECYCLE THINGS RATHER THAN TOSS THEM INTO A DUMP?

We can do that to some extent, and we're beginning to be more serious about recycling. The waste from one process can be useful material for another process. But there are limits to recycling. Remember, the energy used in consumption is dispersed into nature as random vibrations that can't be recovered to do work. Many materials, such as metals, plastic, glass, and paper, can be recycled to make new products by applying more energy to them; it always takes more energy. But eventually those materials wear out and go into the sink. Because expended energy can never be recovered, the scarcity and expense of energy is the main limit to recycling.

SO YOU'RE SAYING THAT RECYCLING, BY ITSELF, WON'T SOLVE THE THROUGHPUT PROBLEM?

Exactly. In nature, all living things eventually decay and become food for other living things. But human-made products either can't be used by living things or are actually poisonous. When they're poisonous, we call them **toxins**. Toxins made by humans have been injuring and killing other creatures, as well as humans, for many decades, because it's exceedingly difficult to isolate them from living creatures. If we bury them, the chemicals will eventually leak out of the dump and get into ground water, lakes, and streams. If we burn them, they may be converted into new toxins as they are emitted into the air. Once in the air or water, they're likely to circulate endlessly from air to water to soil to plants to animals to humans, back to water or air, and around again and again. Remember that we can't really destroy things; they always go somewhere. So we must remember to keep asking, "And then what?"

WELL, IT SEEMS TO ME THAT RATHER THAN STRUGGLE SO HARD TO ISOLATE THEM, WOULDN'T IT BE BETTER TO STOP MAKING THINGS THAT PRODUCE TOXINS?

That's an important possibility. Because we can't make them go away and can't really keep them from injuring us and other creatures, I believe that sooner or later, we'll have to stop making environmentally damaging chemicals. As you can imagine, the chemical industry would violently oppose a proposal to do that. And even if we could be successful in turning away from environmentally harmful chemicals, we'd still have to deal with the toxic load that's already there.

BUT AREN'T THE CHEMICAL COMPANIES LOOKING FOR A TECHNICAL FIX?

Of course they are. I read just the other day that scientists had developed a plasma furnace that can break down chemicals to their original elements. But it takes immense energy to power the furnace. We the educated, especially business leaders, have great faith in technology. But I'm not at all certain that depending on a technical fix will be sufficient. Many technical fixes have unintended consequences that reduce their effectiveness. Furthermore, several thousand new chemicals are developed each year and we allow them to be put to use without serious forethought about their long-range consequences. So it's risky to place all our bets on chemical recycling technology, because it may discourage us from looking for better long-range solutions.

ALLOWING CHEMICALS TO BE USED WITHOUT EXAMINING THEIR POSSIBLE LONG-RANGE CONSEQUENCES SEEMS TO BE A FORM OF BLINDNESS.

Indeed it is. I'm not suggesting that we abandon technology, mind you. We need better technologies to help us conserve energy, reduce pollution, and live more lightly on the earth. I'm only making the point that we can't depend on technology alone to resolve the environmental problems that arise from the way we conduct our affairs.

Here's another example of blindness in the way we conduct our affairs. Even though there are obvious physical limits to both sources and sinks, classical economic thinking and economic measurement pretend that there are no limits. Classical economists measure only flows and ignore what happens to stocks. The Gross National Product (GNP) and Gross Domestic Product (GDP) are only measures of **flows**. When we cut down a forest, economists count that as a gain to the GDP but don't deduct the loss of stock as the forest is decimated. When we pump out the last few barrels of oil, classical economists count only the flow and not the depletion of the stock. Today's GDP calculation is like a car with a speedometer but no fuel gauge. It tells us how fast we're going but doesn't tell us when we're likely to run empty.

Capitalistic thinking depends too much on the ability of markets to solve our socioeconomic problems. Markets work quite effectively to allocate resources, but they can't look ahead to foresee such possibilities as overshoot and dieback, or global warming and climate change. A society that wants to become sustainable must foresee possible scarcities and system disturbances and then use its political system to limit throughput at sustainable levels. Markets also can't distinguish what's morally correct from what's morally reprehensible. They'll respond to the demands for luxury goods by the rich while ignoring the needs of the poor for bare subsistence. For example, land in the tropics that used to grow food for the poor may be diverted to grow flowers for rich Americans who want flowers in the winter.

The free movement of capital and goods in a capitalistic world market has another harmful effect - that of encouraging the location of production facilities in countries with the lowest wages and the lowest levels of environmental protection. Markets encourage, even require, firms to externalize their pollution costs in order to be competitive. The costs then have to be borne by the public and the ecosystem.

Classical economic theorists acknowledge sources and sinks only by consciously excluding them from their theory and calling them **residuals** or **externalities**. The term "residual" means that something is left over and isn't reckoned into economic accounting because it's external to the theory under which the accounts are set up. I'll give you an example of how serious that is: In 1989, a comprehensive study was conducted regarding the externalized costs associated with discovery, shipping, refining, burning, pollution and depletion of crude oil. The researchers estimated that if all those costs had been included, the total cost of a barrel of oil would have been $49 rather than the $18 open-market price of the time.

SO WHO PAYS THE COST NOT REFLECTED IN THE MARKET PRICE?

We all do, through increased taxes and diminished quality of life. And when I say all, I'm including other creatures and future generations. People would be outraged if the price of petrol more than doubled to reflect its true cost. Yet when taxes keep going up to pay for some of the injury and loss caused by the consumption of cheap oil, they scream about high taxes. People in different countries pay different prices depending upon policy objectives that their governments wish to follow. In western Europe, petrol costs two to four times what Americans pay, because governments there try harder to collect enough fuel taxes to cover the costs Americans hide.

Economic accounting makes another huge error when it counts the cost of cleaning up a mess, such as a toxic dump, or a flood, as an addition to rather than a debit from the GDP. Only government regulation has forced firms to incorporate the cost of pollution into their cost accounting. But the cost of controlling and cleaning up pollution is added to the GDP rather than deducted as the loss it really is.

A few economists and governments have finally become concerned about the absurdity of the mistakes in classical economic thinking and accounting and are trying to correct the way we keep our national accounts. The European Parliament convened an international conference on Green Accounting early in May 1995. Americans followed suit with a conference on the same topic in October 1995. But since the United Nations oversees national economic accounting around the world, it may take many years of international negotiation to make the change. In the meantime, we'll have to find other ways to stop deluding ourselves if we hope to continue to live on our planet.

I WISH MORE PEOPLE UNDERSTOOD WHAT'S GOING ON.

The next time someone says we can keep taking from the earth, or keep dumping into it, with no concern for limits, you'll know that you're being misled. The next time someone calls for continual economic growth, you'll know that he or she is calling for a physical impossibility. And the next time you hear about the GNP or GDP, you'll know that you're getting only part of the story.

Chapter 13

Distinguishing Development from Growth

Now let's move on to a related concept. You may have heard the phrase **sustainable development**. It was used a lot at the time of the Earth Summit in Rio in 1992.

YES, I HAVE. BUT I'M A LITTLE VAGUE ABOUT WHAT IT MEANS.

You aren't alone. There's a lot of ambiguity in that phrase, and different people have given it different meanings. Some believe it means that economic growth can be sustained indefinitely into the future - and that's impossible. Before long we must limit our growth, or nature will limit it for us through the collapse of some of its living systems. Nature's solution is very effective. It is death.

It helps if we **distinguish growth from development**. Growth of throughput must stop, but development need not stop. **Sustainable development means learning better ways of enjoying life without injuring the life systems that sustain us and other creatures.**

GIVE ME SOME EXAMPLES OF DEVELOPMENT THAT ISN'T GROWTH.

OK! Take the conversation we've been having. Would you say that it's developed your thinking?

YES, QUITE A LOT, I'D SAY.

We both got more insight by having this conversation, but it hardly affected throughput.

Development without growth characterizes all learning. Not only does it not add to throughput but it's also **not zero sum**. Zero sum is a term economists use for a good that can't be shared. For example, if there's one piece of cake left and we both want it, but you get to it first and eat it, you've won and I've lost. That's zero sum. But if you possess some knowledge that I want and you share it with me, I've gained but you haven't lost. Usually,

we're both enriched from sharing.

To continue, zero sum goods are usually **private goods** that are for the exclusive use of the owner. **Public goods** are available to everybody. A public park is usually not diminished if it's shared. Of course, if too many people want to use it, their enjoyment will be diminished. The air we breathe is a public good; its quality is diminished if people spew too many wastes into it. That's why we must use government to regulate the use of public goods that are subject to overuse.

I make the distinction between public and private goods because development and use of public goods usually add little to throughput but can add a great deal of enjoyment for many people. Public safety is an example of a public good that has deteriorated in developed economies in recent years. Wouldn't it be preferable to develop a better public safety system than to expect everyone to buy guns and be primarily responsible for the safety of his or her home?

I WOULDN'T WANT TO LIVE IN A SOCIETY WHERE EVERYONE WAS ENTIRELY RESPONSIBLE FOR HIS OR HER OWN SAFETY.

As another way of looking at development without growth, think of a singer like Lata Mangeshkar. Her songs have been performed hundreds of thousands of times and have enriched the lives of millions. The fact that they've been shared so widely hasn't diminished their value for any of us; and the performances have added very little to throughput.

I GUESS THAT CONCEPT WOULD APPLY TO WRITING BOOKS, PAINTING PICTURES,

PERFORMING PLAYS, AND MAKING MOVIES.

Publishing books and making movies do add to throughput but the impact is small compared to many other economic activities. A sustainable society probably could afford to do those things.

HOW ABOUT SPORTS?

Participatory sports are healthy, usually fun, and add little to throughput. On the other hand, some energy-intensive sports like car racing and motor rallies use lots of energy to give people thrills; they not only add to throughput but also are polluting. So they should be discouraged.

Notice I didn't advocate spectator sports. If thousands of people come to a stadium or arena for a sports event, and most of them travel by car, they consume lots of energy and create lots of pollution. Spectator sports wouldn't be so polluting if most of the fans used public transportation and if the games weren't played at night. Other sports like hiking, cross-country skiing, sailing, and jogging add only the cost of making the sporting equipment to throughput. They're healthy and can be just as satisfying as sports that use lots of fossil fuels.

Another way to enjoy life without growth is socializing. Many people get great satisfaction from social interaction. Social parties add little to throughput. Civic activity, public policy consultations, and citizen political action are important areas for development that hardly affect throughput. However, festivals such as Diwali, although they promote socializing, devastate city air for days to come. Have you noticed how badly our eyes burn after spending only a short time outside on Diwali night? The effect is clearly noticeable even in air that's already so heavily polluted.

On a more general level, we make a fundamental mistake when we think we have to consume lots of material goods to have a good life. As Alan Durning said, "If we attempt to preserve the consumer economy indefinitely, ecological forces will dismantle it savagely. If we proceed to dismantle it gradually ourselves, we will have the opportunity of replacing it with a low consumption economy that can endure. As long as we limit our population and our throughput, we can continue to develop and to enjoy life without injuring life systems.

CAN WE EXPECT THIRD WORLD COUNTRIES LIKE INDIA TO DEVELOP WITHOUT GROWING?

Because many of them are desperately poor, they need to increase their throughput somewhat. Most Third World leaders are very eager to develop and usually believe they can follow the path of development chosen by the now industrialized. But the inherent limits on sources and sinks in the earth's life systems won't allow them to. People in the developed world should be more sympathetic and try to help us with modest development in a wise way. We can't become as wasteful as they are; and they, in turn, have to change their wasteful ways. If they in the developed North could reduce their consumption by 20 percent, the developing South could grow for a while to improve its basic subsistence without adding a huge burden to life systems. But these proposals would only work if the world simultaneously restricts population growth severely.

BUT WOULDN'T THESE PROPOSALS BE DIFFICULT TO PUT INTO PRACTICE?

Most people in less developed countries (LDCs) want to be rich and wouldn't be content with mere subsistence. And most people in more developed countries (MDCs) are having difficulty making ends meet now and would resist the idea of giving 20 percent of their wealth to the LDCs. Also, many millions of people strongly reject the idea of limiting the size of their families.

AND THEN WHAT?

I'm afraid that everyone will continue trying every way possible to grow economically. Since Earth systems can't support economic and population growth for very long, growth will stop anyway because of system failure. That will bring on painful social learning that may or may not be effective.

AND THEN WHAT?

Many people will die, but life will go on. Since people starving to death will scrabble to stay alive even for one more day, the earth's life systems are likely to be reduced in capacity. The survivors are likely to have a new opportunity to build a sustainable society. Because of the painful social learning brought on by system failure, maybe they'll have a chance to succeed the next time around.

I FEAR FOR THE FUTURE OF MY GRANDCHILDREN. WE SHOULD DO EVERYTHING WE CAN TO PREPARE THEM FOR THE DIFFICULT DAYS AHEAD.

Chapter 14

Distinguishing Sustainability from Sustainable Development

IS SUSTAINABLE DEVELOPMENT THE SAME THING AS SUSTAINABILITY? IF NOT, CAN WE DISCUSS WHAT SUSTAINABILITY MEANS?

They're related but different. To explain the difference, we'll have to go back over some points we talked about earlier.

THAT WILL HELP ME PULL THINGS TOGETHER.

The term **sustainable development** applies to activities in the human system, whereas **sustainability** is important for biogeochemical systems too. When a system is sustainable, it will keep functioning for the indefinite future. Long ago, nature learned how to make its systems sustainable. If humans weren't here, it would soon restore its systems to sustainability. Human systems that aren't sustainable can't provide a good life for very long. **A society trying to be sustainable must maintain the integrity of life systems. A contemplated action is right when it tends to maintain the ecological**

diversity, integrity, and sustainability of Earth's life systems, and wrong when it tends otherwise. If we act against that precept, we run a serious danger of diminishing life support for ourselves and other creatures.

HAS THIS ALWAYS BEEN THE CASE?

That is a very important question. For thousands of years, humans did foolish and terrible things to nature, but the earth always seemed to heal the wounds and life continued much as before. We came to think of life systems as resilient to our abuses. We came to think that the future will be like the past, and most people still think so today. But a key point many environmentalists make is that **we can no longer predict the future from the past**.

SO WHAT HAS CHANGED?

As you know, human population is booming and still growing fast. That's happened because we have acquired more and more power to influence and change physical processes. We got better and better at doing what we've always been told is praiseworthy, such as having children, conquering disease, living longer, dominating nature, acquiring more goods, commanding energy, traveling the world, and consuming enthusiastically. Without intending to, we polluted our air, water, and soil; we expanded deserts, deforested millions of acres, made the rain acidic, thinned the ozone layer, depleted resource stocks, and changed climate patterns. And now we're on the verge of driving life systems into chaos. We now have the capability to destabilize life systems to the point where they'll no longer support life as we know it. Even if humanity learns a bitter lesson and tries to live harmoniously with nature, it could take centuries for nature to regenerate the viability the system had before human disruption.

WHY CAN'T WE LEARN AND CHANGE BEFORE WE DRIVE LIFE SYSTEMS INTO CHAOS?

I wish we could, but it's clear to me that we haven't altered our ways of thinking to stay in tune with the new reality we've created. We still fail to look ahead; we still think the future will be like the past; we haven't learned to think systemically; and we still put our own needs above the needs of life systems and society. A global society so populous, so powerful, and so blind is not sustainable. If we won't learn and won't adapt, we won't endure. Change will come whether we foresee it or not and whether we want it or not.

Chapter 15

Choosing Wise Policies

YOU KNOW, THERE ARE LOTS OF PEOPLE, SOME OF THEM SCIENTISTS, WHO CLAIM THAT WHAT YOU ENVIRONMENTALISTS ARE SAYING IS NONSENSE; THAT MANY SCARY PREDICTIONS ABOUT THE ENVIRONMENT HAVEN'T COME TO PASS; THAT THERE'S NO SOLID EVIDENCE THAT THINGS ARE GETTING WORSE.

The public has a very high regard for science because it seems to work magic and do powerful or ingenious things that would have been impossible in earlier times. It's widely believed that science can do nearly anything it sets its mind to, if given enough time and money. Most people also believe that science is very exact and, because it knows things with such precision, it's not to be doubted.

For those reasons, most people who are involved in debates about how the world works, or who are pushing a particular policy, back up their proposals with scientific findings or try to get authoritative and prestigious scientists to support their side. Since all sides in an argument usually can find a scientist or two to support their perspective, the media frequently present conflicting scientific opinions.

Scientists become weapons to be used in political battle rather than impartial observers. Consequently, some people are wary of accepting opinions solely because they come from a scientist.

When scientists disagree, we should keep in mind that uncertainties are normal in science because it's always open for revision. Scientists typically argue vigorously about the meaning of the available evidence, so it takes some time for a scientific interpretation to be "settled", and even then it may be overturned. The history of science shows many instances where a widely accepted perspective is replaced by an alternative perspective because countervailing evidence persuaded most experts to change their minds. Knowing what reputable scientists believe about the way the world works helps us a great deal in understanding the world, but scientists' beliefs aren't an infallible guide. Even if scientists know something with certainty, they may not know how best to apply that knowledge to preserve the viability of life systems.

Knowing what to believe becomes very problematic when we're trying to foresee the future. We can never know the future with certainty. Whenever we make policies, we're guessing what the future will be like. Environmentalists often advocate that we **learn to think probabilistically**; that is, choose policies that are most likely to be wise. Science isn't black and white and policy-making can't be that way either. Even physics, the most exact of the sciences, had to turn to probabilistic thinking when its practitioners tried to understand phenomena at the subatomic level.

Knowing the proportion of expert scientists in a field who agree with a certain position is a useful clue for deciding which policy is most likely to be wise. If 90 percent agree with position A while 10 percent take position B, position A is probably the better policy.

Another intelligent strategy is to choose a proposed policy that's closest to the way nature does things. Evolution has no brain and can't plan; it doesn't know where it's going, but it's an extremely patient and effective editor. Evolutionary development has been making mistakes and correcting them for millions of years. What survives is what fits best with nature's grand processes. Humans are most likely to survive and flourish if they **make choices that accord with the way nature works**.

CAN YOU GIVE ME AN EXAMPLE?

Remember the story I told you earlier about the settlers in Australia who introduced rabbits to the Australian ecosystem so "gentlemen could have sport"? They unleashed a cascading set of consequences that nearly destroyed the ecosystem. If they had better understood th importance of fitting into the existing ecosystem, they could've avoided lots of grief.

We faced a similar dilemma in choosing how to deal with insects that threaten farms and gardens. Chemists developed poisons, unknown to nature, to kill these pests. Humans eagerly adopted the powerful insecticides without asking, "And then what?" Now, millions of tons of chemical poisons are applied all over the planet, creating dire unforeseen consequences: the insecticides kill many beneficial insects; the poisons circulate in our soil, water, and air, bringing disease and injury to many other creatures, including ourselves; the poisons enter the food chain and end up in our own food; and insect pests develop immunity to the formerly lethal poison and return in force to plague us.

People in the chemical industry who depend on the continued spreading of chemicals for their livelihood fiercely resist efforts to stop using these poisons. Their scientists assure us that the chemicals are safe and effective and encourage consumers to use them with reckless abandon. Other scientists now recommend integrated pest management that uses natural predators of undesirable insects and poisons already well-known in nature. How should we choose? The path that is more likely sustainable is the one closest to the way nature works. Did those examples help?

YES, FOLLOWING NATURE'S WAY SEEMS TO BE A GOOD GENERAL POLICY.

There's yet another way to help decide which course is probably best. We should always ask, **"What is the cost of being wrong?"** Philosophers call this dilemma **"Pascal's Wager"**. The French philosopher Pascal couldn't quite decide whether or not there is a God. He reasoned that if he wagered that there is a God, and lived a virtuous life, but later found out there is no God, he wouldn't have lost much, because his life would still have been pretty good. On the other hand, if he wagered that there is no God, and lived a sinful life, but later found out that there really is a God, he'd be condemned to hell. So he acted cautiously and bet God exists.

The U.S. government reasoned similarly when it confronted the threat of the Soviet Union during the Cold

War. It was never sure the Soviet Union would attack, but it believed it was acting prudently when it spent trillions of dollars to prepare the country to defend against an attack.

Similarly, we can't be absolutely certain that the climate system will become persistently overheated or severely chaotic if we stay on our current trajectory. But there's a good chance that it will. Consequently, it would be extremely foolhardy to rule out the possibility. So wouldn't it be better to take precautions to avoid ever finding out if chaos will result? Environmentalists and diplomats call this position the **no regrets policy**. Later, if better science shows us that climatic chaos wouldn't have occurred, we'll have lost little. But if we refuse to change and our climate becomes severely chaotic, we'd lose a great deal - perhaps our society and our lives. Choosing to stay the course can be just as dangerous or more dangerous than changing course. Those who resist acting until science can be certain that climate change will occur expect more of science than it can deliver, and choose a path that's more likely to be calamitous.

BUT SUPPOSE WE DON'T CHANGE, AND WE DO IN FACT DRIVE THE CLIMATE TO BEHAVE CHAOTICALLY., AS YOU JUST PUT IT. WHAT HAPPENS THEN?

I can only guess what might happen, because humanity hasn't traveled this road before. But I can imagine two possible scenarios. In the first scenario, frequent and unexpected climatic disasters (floods, droughts, hurricanes, forest fires, and tornadoes) are interspersed into "normal" climate patterns. Thousands, even millions, of people could die, and the property loss could be devastating.

Loss of the premise of continuity would be the most serious consequence. We all implicitly assume that tomorrow will be like yesterday; it's a crucial anchoring for our lives. For example, economic thinking abhors uncertainty about the future. If people don't have confidence that the future will be like the past, investors won't invest, lenders won't lend, people won't build houses, entrepreneurs won't start businesses, and young people won't prepare for careers. Confidence that climate patterns will be similar to the way they were in the past is the most fundamental of all premises of continuity.

Recently, the world's large insurance companies have become seriously concerned about the possible consequences of climate change. Their entire business is premised on continuity. They now realize that the nature of their business puts them on the front lines of the climate problem, because they'll be expected to absorb the financial shocks of weather-related disasters. They'll probably fight the fossil fuel industry, which has been urging the delay of precautionary actions until there's absolute certainty about the pace and magnitude of climate change. Waiting for consensus that climate change is coming could destroy the insurance industry.

Notice that I'm talking about climate, not weather. We all know that weather changes from day to day; but climate pretty much has followed the same pattern from year to year. If changes in climate behavior lead people to doubt that Earth's natural systems will behave as they have in the past, economic activity is bound to decline.

The economic growth that all governments currently covet can't happen in a mood of great uncertainty.

In the second possible scenario, Earth might experience an extended period, maybe a decade or two, of oscillation-type chaos in the climate system. Plant life would be especially vulnerable to climatic oscillation because plants are tuned to the normal climate pattern of their niche and can't readily migrate to more favorable niches if the climate changes. Plants are injured or die when the climate is too hot or too cold or too dry or too wet. Since plants are the source of food for all other creatures, plant dieback would lead to a severe drop in agricultural production. Farm animals and wildlife would die in large numbers. Many humans would starve. Several years of climatic oscillation could result in the deaths of 2 to 4 billion people.

World financial markets would collapse in a chaotic situation. Their collapse would, in turn, lead to sharp declines in commodity markets, in world trade, in output from factories, in retail sales, in research and development, in tax income for governments, and in education. Non-essential activities such as tourism, travel, hotels, restaurants, entertainment, and fashion would also fail. Billions of unemployed humans would reduce their consumption drastically and focus on merely subsisting. Our economic system would collapse like the house of cards it really is.

THAT'S PRETTY SCARY. BUT PEOPLE WILL PUT IT OUT OF THEIR MINDS, BECAUSE IT WON'T HAPPEN FOR A CENTURY OR MORE. THEY AND THEIR CHILDREN WILL BE DEAD BY THEN.

Be cautious about predicting the future by the past. This change could come much sooner than you think. Unless we take preçautions, it's likely to affect our children and grandchildren. That's why it's so urgent to reach people now.

Chapter 16

Theories of Social Change

You just said collapse wouldn't happen for a century or more. How do you know that?

EVERYBODY KNOWS BIG CHANGE TAKES A LONG TIME.

Your answer illustrates a very important fact. **Everybody has a theory about how social change works.** Similarly, we all have a theory about social change. Actually, the two types of theories are related, because what we believe about the world influences what we believe is possible with respect to social change.

HOW CAN YOU BE SO CERTAIN THAT EVERYONE HAS A THEORY OF SOCIAL CHANGE? I KNOW SOME PEOPLE WHO DON'T SEEM TO HAVE ANY THEORIES AT ALL.

A theory doesn't have to be structured abstractly and be elaborate to be a theory. A smattering of conventional sayings can serve as a theory a person lives by. Every time we choose a course of action we're implicitly using a theory of social change. Usually our theory about social change involves basic beliefs about human nature.

For example, the other day I listened to a public official explain why he believed a new law should contain strong penalties. He said, "Without pain, people don't learn". He was so certain about his theory that he stated it as a fact. In a similar vein, an M.P. said, "Unless we have a nuclear bomb, they won't change their policy". Many actions in everyday life are based on just such snippets of theories about social change. And most disagreements in legislative bodies are rooted in differing theories about human nature rather than in different desired outcomes for society.

BUT WHERE DO THEORIES OF SOCIAL CHANGE COME FROM?

Most of us learn them early in life from the conventional beliefs of our culture. As with theories about how the world works, theories about social change are deeply embedded in our belief system. Most of us aren't conscious of our premises about human nature and social change; we're not aware of how much they influence the way we think. That's one of the reasons legislative bodies get tied up in gridlock. Each side is firmly convinced that it knows how human nature works, and that its policy is best.

BUT WHY DO YOU MAKE THAT POINT?

Because when we confront a problem and try to decide on a course of action to correct it, our theory of social change unconsciously comes into play. We may be unaware of how it restricts and misguides our thinking. To liberate ourselves, we can consciously examine our theory of social change for its validity. Unless we do, we may not be able to discover why many of our actions fail, or understand why someone disagrees so strongly with an argument we firmly believe is correct. By reexamining our social change theory from time to time, we become conscious of knowing how we know. Knowing ourselves that well liberates us to explore problems more deeply.

Duane Elgin has developed an elaborate theory of culture and consciousness in which he invites us to think about eight stages of consciousness development. According to his theory, most people in contemporary society are at the fourth stage which he calls **dynamic consciousness**. He predicts that our next stage of evolution will be an era in which most people develop **reflective consciousness**. People with dynamic consciousness are conscious of their thinking. Creative thought and material development give meaning to their lives. But people who evolve further into reflective consciousness become conscious of their consciousness. They're able to stand back and observe the ego self and be both observer and observed. People who become critically aware of how they developed their beliefs about how the world works, and their beliefs about social change, are developing reflective consciousness.

THAT'S AWFULLY ABSTRACT; CAN YOU GIVE ME AN EXAMPLE?

I grew up in a rural community and unconsciously acquired a smattering of sayings about social change from the community's conventional wisdom. When I went off to college, I used to tell my friends that "planning doesn't pay". A college friend startled me by challenging the validity of the saying (something that wouldn't have happened in my hometown). For the first time in my life I thought seriously about that saying. I wondered where it came from and was annoyed with myself for using it without reflecting on its meaning.

When farmers plant crops, they have a simple plan that the crops can be brought to harvest and that they'll be rewarded with food for their families and a little saving. But weather (too dry, too wet, too hot, hail storms, and so forth) can wreck their plans. So in the rural context of weather-dependent farming, the saying "planning doesn't pay" has some validity. But it's not equally valid in other contexts. Developing reflective consciousness allowed me to see that social change must be planned for and understood in context.

A tussle between these two kinds of consciousness is currently at work as we as a country contemplate how we can avert global warming and climate change or how we can change our society from its current unsustainable state to a sustainable state. It's not hard to imagine the changes needed to make society sustainable, but it's immensely more difficult to imagine how millions of people can be moved through massive transformations in thinking and behavior to actually bring about a sustainable society. Our dynamic consciousness looks at where people are now and then at where they must go and concludes that the task is hopeless. Never in the past has change of that magnitude been achieved.

But when we look at the same problem with reflective consciousness, our perspective broadens. We can see that the blow that made the dinosaurs extinct was far more damaging to life systems than chaos in climate systems would be. Yet life systems recovered, evolved, and flourished even more splendidly than before. Reflective consciousness also lets us see more clearly that, since an unsustainable society cannot persist, social change will come whether it's foreseen or not; it's inevitable that the social context will change. In the present context of modern society, we'll have, for a short time, the option of staying the course. Before long, though, that option will no longer be available, because nature's forces will change our course for us. In the new context, our collective mind will be able to contemplate transformations that are now unthinkable. When we use reflective consciousness to imagine ourselves in that new context, societal transformation is not only reachable, it's inevitable.

Let me get back to your point that big changes take a long time. If we examine that premise critically in the light of history, we observe that most big changes did take place over a very long time. But we also discover that sometimes big change comes astonishingly swiftly, as shown recently by the breakup of the Soviet Union. We can also see that the pace of change is speeding up. Elgin makes the point that it took humans about 50,000 years to evolve out of the hunter-gatherer era into the agricultural era, and 5,000 years to evolve out of the agricultural era into the industrial era. But we're already evolving out of the industrial era after only 300 years. Humans are speeding up the pace of change because of our huge numbers, our intelligence, and our power. We're changing global biogeochemical systems, economic systems, political systems, and learning systems. And our growing recognition that we must change the way we do things will further accelerate change. Those are all important reasons why we should be cautious about predicting that the future will be like the past. The safest prediction is that change will come whether we want it to or not, and that it will probably be faster than we expect.

Chapter 17

Next Steps

I SEE YOUR POINT THAT CHANGE IS COMING WHETHER WE WANT IT TO OR NOT. BUT HOW DO WE STEER IT SO THAT SOCIETY MOVES IN A MORE SUSTAINABLE DIRECTION? HOW DO WE AVOID GETTING INTO THE CHAOTIC SITUATION THAT YOU DESCRIBED?

That's the big question, and we'd all like to know the answer to it. To answer it, I must use my theory of social change, and both you and I should keep in mind that it's a theory that should be examined critically. I've been examining it critically for many years, and it's as well thought through as I can now make it. I've given you part of my theory in what I just said. But to answer your specific question about avoiding chaos in the climate system, we have to look at the situation of people right now.

My theory begins with the premise that people cannot change if they're not paying attention. The greatest difficulty is to wake people up, to get them to pay serious attention to a problem on the horizon while they're busy with other problems right in their lap. Trying to scare them with the chaotic scenario I just described probably wouldn't be effective. My theory says that people have a powerful tendency to deny messages they don't want to hear. So they'll more likely believe people who say there's no need for preventive action. Feeling they have more than enough worries now, most people will try to enjoy a few daily pleasures and not worry about possible trouble in the future.

The stark reality, then, is that most people aren't listening. They don't know what's coming and they don't want to be bothered. It'll probably take a very strong jolt from nature before they'll pay attention.

DO YOU MEAN A JOLT LIKE A POWERFUL STORM?

It'll probably have to be even more powerful. Have you noticed how people rebuild their homes in vulnerable places even after they've been demolished by floods, hurricanes, or earthquakes?

YES, THAT'S TRUE.

A jolt that's large enough to change them will probably be a collapse of natural systems that they've always depended on. It'll have to upset nearly everyone drastically. Furthermore, most people will have to see clearly that they no longer have the option of staying the course. They'll have to experience the loss personally. Merely describing the expected loss is unlikely to wake them up. Even if most people know for sure what lies ahead, some will still resist changing.

WHY CAN'T PEOPLE LISTEN NOW, SO WE DON'T HAVE TO GO THROUGH THE PAINFUL LOSS THAT'S LIKELY? THERE MUST BE SOME WAY TO GET THEM TO LISTEN.

Other than what I just said, I don't understand why people refuse to listen. I think about that question every day.

I also know that getting millions of people to think about a problem has great potential for creativity. Maybe you, or people like you, can come up with a better idea for reaching people with this difficult message.

NOTHING OCCURS TO ME RIGHT NOW. BUT SUPPOSE SOMEONE COMES UP WITH A GOOD IDEA FOR REACHING PEOPLE. WE'D NEED A CONCISE LIST OF WHAT NEEDS TO BE DONE TO AVERT GLOBAL CALAMITY.

Well, **if** I could wave a magic wand to get people and their governments to cooperate, the following thoughts and actions would be effective:

- Recognize that the earth, and the continued vitality of its life systems, is the urgent reality to which our civilization must give primary attention.
- Love creation, love the community of life, love Earth. Prolonged, meaningful, and intimate contact with nature is so enriching that it develops deep and lasting love.
- Affirm love, or caring for others, as a primary value. Love not only those near and dear but those in other lands, future generations, and other species.
- Re-evaluate daily activities in order to live lightly on the earth.

- Change the way we think as quickly as possible. We need to clarify our values and adopt new priorities. In the process, we should clarify our responsibilities, so that people see their part of the overall task as well as accept the necessity to do their share. All of us must learn to think systemically, holistically, integratively, and in a futures mode. We should also strive for reflective consciousness. Renewed reflection on the true meaning of quality in living should be part of this relearning.
- Think globally, act locally.
- Think tomorrow, act today.
- Control and gradually eliminate weapons of mass destruction.
- Stop population growth as quickly as possible. With heroic efforts, world population might be leveled off at 8 or 9 billion people. Choose small families.
- Reduce material consumption in the more developed countries and use that reserve to help the less developed countries meet their basic needs. A sustainable, peaceful world of the future must have only limited inequality.
- Cut back as much as possible on use of fossil energy; develop and adopt more energy-efficient technology; cut out all energy waste; stop using fossil energy simply for thrills, fun, ease, or comfort; convert to solar energy.

- Aggressively reduce economic throughput to preserve more resources for future generations and to reduce the discharge of wastes into the biosphere. Failure to do this will seriously reduce the carrying capacity of life systems.

- Find ways to share employment so we don't need to make unneeded goods just to provide jobs for people. Work should be redefined to become a means of self-realization, not just a pawn in economic competition.

- Emphasize making quality products that can last lifetimes, beautiful things to be cherished and preserved. Products should be designed to be easily repaired and safely disposed of. They should be marketed with as little packaging as possible.

- Diligently reuse, restore, and recycle materials that we now throw away. Carefully dispose of all other wastes.

- Eliminate use of chlorofluorocarbons (CFCs) so the stratospheric ozone layer can restore itself. Recapture CFCs from current uses and break them down into harmless compounds.

- Protect the purity of the water and the air; restore their purity where it's degraded.

- Conserve top soil and rejuvenate degraded soils.

- Phase out energy- and chemical-intensive agriculture so we can develop sustainable methods of tillage and crop production.

- Stop the release of toxic chemicals into the environment.
- Restore degraded ecosystems to health.
- Protect and enhance biodiversity.
- Conserve nature and resources so future generations and other creatures can enjoy a life of decent quality.
- Plant a tree. Earth needs billions more trees.
- Diminish the rewards for power, competitiveness, and domination over others. A sustainable society emphasizes partnership rather than domination, cooperation more than competition, justice more than power.
- Develop a procedure in the private sector, with government supervision, for careful review and forethought to estimate the long-term impact of a proposed technology. Bad consequences of new technology are easier to avoid or manage if they're anticipated from the start.
- Redesign government to maximize its ability to learn; then use the governmental learning process to promote social learning.
- Develop a new governmental institution to better anticipate the future consequences of proposed policies, laws, and technologies.
- Make the learning of environmental thinking a national and global project. Provide every child with an environmental education - it's just as basic

as history. Establish environmental education programs for adults. Make a special effort to educate media employees about environmental concepts and thinking.

- Expand opportunities for people to experience and participate in environmentally sensitive art, such as plays, songs, paintings, and poems that illustrate the good and bad ways in which we relate to nature.
- Don't merely work for a living, but work for something that's truly important.
- Keep a sense of humour, sing, dance, affirm love, be joyous in your oneness with the earth.

THAT'S A LONGER LIST THAN I EXPECTED. I'M GLAD YOU INCLUDED THAT LAST POINT. IT'S SO EASY TO GET DISCOURAGED WHEN FACING SUCH HUGE PROBLEMS.

If we thought some more, we could add to that list and expand on many of the suggestions. The list demonstrates that a sustainable society would be quite different from the society we have now. Getting there will require the creative thinking of millions of people. The new way of thinking I've proposed is a good beginning.

ARE THERE ANY EXAMPLES OF SUSTAINABLE SOCIETIES ON EARTH THAT COULD SERVE AS A MODEL OF THE SOCIETY WE SHOULD TRY TO BECOME?

As you can imagine, environmental scholars

have been searching for such a model. The one that comes closest to being sustainable in a long-run harmonious relationship with nature is **Kerala**, a state in southwest India. It has a high literacy rate, a low birth rate, low infant mortality, and low throughput. It uses lots of surplus human time and energy to work with nature and help natural systems renew themselves. The caste system carries much less importance with the people there. Women have nearly equal status with men and both participate regularly in community affairs and politics. So far as we know, no other society has this particular combination of characteristics.

BEING AN AMERICAN, I AM CURIOUS TO KNOW HOW YOU THINK THE UNITED STATES, BEING ONE OF THE MOST POWERFUL AND INFLUENTIAL COUNTRIES IN THE WORLD, COMPARES WITH OTHER COUNTRIES AS FAR AS BEING READY TO BECOME A SUSTAINABLE SOCIETY.

To be blunt, Canada and the United States are the most unsustainable societies on Earth. Their energy consumption per capita is the highest in the world. They refuse to cut back on fossil fuel consumption, which is the single most critical action for avoiding chaos in our climate. And their consumption of other goods is equally wasteful. To put it another way, they have the most to unlearn before they can achieve a sustainable lifestyle and a sustainable society. I believe the necessary learning will be quite painful. In the meantime, we have to avoid learning from them what they themselves need to unlearn.

THAT'S BAD NEWS, BUT I HAVE TO ADMIT YOU'RE PROBABLY RIGHT. WHEN I THINK

ABOUT THE IMMENSE CHANGES THAT MUST TAKE PLACE, I FEEL OVERWHELMED. I'M JUST AN ORDINARY PERSON, AND IT SEEMS THERE'S ALMOST NOTHING I CAN DO. WHY SHOULDN'T I FORGET THE WHOLE THING?

I feel that way too, almost every day, actually. But then I remember how swiftly some changes have come about that no one expected could happen. The Americans changed their ideas about smoking in only two decades. The countries in Eastern Europe withdrew from the communist system in about three months, and, in most cases, without any bloodshed. The Soviet Union disintegrated with hardly any military action. Those changes weren't forced from without; they took place within people's minds. New ideas seemed to spread from mind to mind almost like wildfire. People were able to use their own learning to produce change by adopting new ideas. The time for new ideas just seemed to come.

May be the time isn't far away that people can hear the ideas we've been discussing. You can help by changing your own lifestyle to a more sustainable one and by spreading these ideas.

I'M WONDERING IF YOU COULD SUMMARIZE THE ESSENCE OF WHAT YOU ARE ARGUING FOR.

I recently gave a seminar on environmental thinking at Cornell University, U.S.A. As a handout (Fig. 17.1) for the participants, I prepared a one-page summary contrasting environmental thinking with conventional thinking. You might use it as a reminder of the main features of environmental thought.

Figure 17.1 : Characteristics of Environmental Thinking

	Environmental Thinking replaces	*Conventional Thinking*
1	holistic, systems embedded in systems, everything connected to everything else	parts equal the whole, minute analysis, anatomical
2	systemic, multiple causes and feedbacks, focus on processes, relationships, and networks, we can never do merely one thing	linear, mechanistic, focus on immediate cause and effect, unaware of multiple consequences
3	integrative, stability in diversity & complexity complicated to suit reality easy grasp	no search for interconnections unconcern with diversity, over simplifies the complex - seeks
4	future - oriented, perspective from far in past to far in future, wary of predicting future from the past, always ask,"And then what?" alert and anticipatory	present - oriented, one day at a time, assumes future will be like the past, unconcerned, indifferent
5	probabilistic thinking	searches for exactness & absolutes
6	reflective consciousness, conscious of knowing how one knows	dynamic consciousness, conscious of thinking
7	recognizes 1st law of thermodynamics, recognizes there is no "away", everything has to go somewhere	unaware of 1st law of thermodynamics, throwaway mentality
8	limits to throughput and population growth, distinguishes growth from development, wary of growth, concern about doubling times, wary of tragedy of the commons, wary of overshoot & dieback	denial of limits, confuses growth with development virtual worship of growth, unaware of doubling times, unaware of impending tragedy, unconcerned with limits
9	concern about effects of entropy	unawareness of entropy
10	distinguishes public & private goods	confuses public & private goods
11	concern about loss of ozone layer	unaware about threat of UV-B radiation
12	concern about greenhouse effect	unaware of threat of climate change
13	concern about loss of ecosystem	unaware of threats to ecosystem integrity

The heart of my plea is that we must stop seeing nature and life systems as separate from society. **Nature, life systems, and society are one; they have a common fate.** Humans can continue as part of our magnificent natural heritage only if they live in harmony with Earth's life systems. When you come right down to it, **we have no other choice**. Learning how to live in harmony with natural systems must be our grand global project.

THANKS FOR BEING SO WILLING TO SHARE. I'LL TRY TO BECOME A BETTER CITIZEN OF PLANET EARTH.

Thanks for listening. I wish everybody was as eager to learn.

Afterword

When I received a complimentary copy of his new book *"LEARNING TO THINK ENVIRON-MENTALLY While There is Still Time"* from my friend Prof. Lester Milbrath from Buffalo, New York, U.S.A., warmly written

"TO MADAN (DEWAN)

In appreciation for your continuing good work for a better future for all creatures

LES MILBRATH"

I went through it with great interest. I found the style and presentation unusual and appealing. I asked Les's permission to get this book published in India - its Indianized version - and spread it to schools and colleges and others interested in the Environment. Fortunately, one of my young volunteer friends, Ravi Gulati, was with me when I received the book and he kindly offered to get an Indianized version ready plus some adapted illustrations to make the book more interesting for the Indian audience.

Although this has taken some time, it is ready now and I do appreciate the Indianization efforts by Ravi.

I am supposed to write some introduction. I felt like

A) Modifying the name to LEARNING TO THINK AND ACT ENVIRONMENTALLY While There is Still Time (hardly).

B) Calling what I am writing as an **"Afterword"** and not Introduction or Foreword.

C) Adding some tips on Action and including an experience of Delhi and India, including a summary of two publications that we have:

– 91 Ways you can help save our Environment

– Towards Cleaner Delhi Cleaner India.

This is being done now. In reflecting and analyzing the new title **"LEARNING TO THINK AND ACT ENVIRONMENTALLY - While There is Still Time"**, the learning process is Self reflection, but also reflection over the experiences of "Cleaner Delhi Training Centre" which is working for the last 36 months, where I have worked as Director for 1 year and now as Chairman of the Training Centre. The Indian Institute of Ecology and Environment as well as the Rotary Club of Delhi South Metropolitan and Rotary District 3010 have given great support to this Training Centre and programme. Over 300 students joined in the course. Highlights of the Training Course include all aspects of Air, Water and Soil Pollution etc. The session on 2nd October, 1996 included the following:

1. Garbage Management
2. Demonstration of Vermi composting
3. Brainstorming session with many participants including school children and teachers

4. Suggestions and Recommendations

   i. Starting a newsletter on the CLEANER DELHI ACTION PROGRAMME

   ii. Adoption of a village/slum by school children for improving the environment of the area

   iii. Co-operation with M.C.D. regarding garbage collection at specific points in the colony/city to ensure cleanliness and thus reduce/remove the problem of garbage which leads us to liver and other diseases

5. Consortia for Cleaner Delhi, and for each Cleaner City

The results are still much below expectation and TIME IS RUNNING OUT. We have, however, not lost hope and are determined to continue the work with young people - school and college students for whom this book is specially meant. A large number of students have signed a pledge as follows:

PLEDGE

"Recognizing the potential of the younger generation, I undertake to dedicate myself to the cause of `Cleaner Delhi Cleaner India', and to make aware at least 2 other young people to spread the message and action for a Cleaner Delhi, Cleaner India."

However there has to be more political will and more legal Action (Thanks to the Judiciary which is making Government/people more alert). The path is very difficult, arduous and thorny, but many young and old people have to dedicate their lives to the cause of Cleaner Cities, Cleaner India, Cleaner World. I am reminded of Nobel Laureate, Poet Rabindranath Tagore's advice in "Stray Birds" First Edition 1917.

(No. 56) *Life is Given to us*

We Earn it by Giving it

(No. 301) *Let Me Feel This World*

As Thy Love My Lord Taking Form

Then My Love And Service

Will Help It (Underlining mine)

As regards acting Environmentally, ie, for a better Environment, we have organized 91 Actions into 11 convenient sub-headings of environment. These are :

1. Water
2. Electricity
3. Gas
4. Sanitation and Sewage
5. Travel and Transportation
6. Recycling
7. Environment friendly technologies
8. Himalayan Conservation
9. Planting trees
10. What our community leaders can do specifically?

11. Healing the Earth, healing ourselves

As students, as community leaders, as Rotarians, we can all help save the Environment effectively. Simply stated, we have to practise what we preach. People are likely to pay more attention to what we are doing than what we are saying.

As the community leaders, we have to act as an Environmental Watch. Create or be part of a pressure group for local, regional, national or international Environmental issues.

Increase awareness of ecological concerns and initiate projects relating to improving the environment.

Charity begins at home. Begin with the immediate environment.

Water

- Fix leaking taps. 10 drops per minute waste about 3000 litres of water a year.
- 11 litres of water is used every time you flush the toilet. Please do not flush unnecessarily. If you are fixing new toilets, fix *Cascade.* You can control the flush.
- In gardens, farms and elsewhere, use water wisely and efficiently.
- India is one of the richest water resource countries. Still we often have droughts and floods. Help in projects and programmes to save water; harvest rain water and use the water more efficiently and wisely.

Electricity

- Use energy saving CFLs (Compact Fluorescent Lamps) and tube lights rather than incandescent bulbs, specially in kitchens where lights are kept on for a long time.
- Put off lights when not in use.
- Install fans instead of costly air conditioners. Where air conditioners are installed, use them only when necessary. Keep blinds and curtains drawn and doors closed to reduce use of air conditioners.
- Make maximum use of natural light. Think about it. Do you really need to turn on a light during the day?
- Use Solar Water Heater where possible to replace electric geyser.

Gas

- Cover the saucepans. It costs 4 times as much to keep contents of uncovered saucepans at boiling point.
- Use pressure cookers as far as possible. It saves considerable gas.
- There are conditions where use of Solar Cooker can help save gas.

Sanitation and Sewage

- Encourage authorities to have suitable well cleaned urinals for the public at suitable places. Discourage the public to use the roadside as urinals.

- Promote installation of Okhla type sewage plants for bio gas and recycling.

Travel and Transportation

- Try forming a car pool. Traveling with your friends or work mates will cut down traveling costs. A car pool can be cheap if 2 or more ride as a group.
- Synchronized traffic lights cut down on fuel waste.
- Shop nearer home. Do not drive miles to save a few rupees.
- Do you really need to drive a car everywhere? Walk to work, or ride a bicycle.

Recycling

- Develop a recycling attitude. Many households give newspapers, cans, glasses, etc for recycling. That system should become an attitude and a habit.
- Buy fresh products from your local market rather than over packed supermarket goods thus choosing goods which have the least packaging. That means less wastage of energy and resources, less garbage and general saving for all.
- Build a compost pit in your yard for grass, leaves, branches and other clippings. It will reduce waste collection and help your garden grow.
- Eggshells, banana peels and vegetable and fruit peels can go into the compost heap.
- Start a recycling programme at the office or home. Cans, glass and newspapers are easy. Many

companies can also buy recyclable bond and computer paper.

- Through your school board or parent teacher group, encourage local schools to teach ecology and recycling as part of the curriculum.

Environment friendly technologies

- Air pollution control equipment; air scrubbers; the development of a clean car engine and also control quality of exhaust gases.
- Effective use of fly-ash bricks (made from the waste from power generation industry) in the building industry.
- Garbage compactors, waste incinerators, hazardous waste management technologies.
- Waste water treatment in the pulp and paper industry. Chromium recycling in tanneries and reactors for recovering heavy metals from industrial waste water.

Himalayan Conservation

- Himalayas being the life-line and in a stage of degradation, each one of us can help even in a small way to save Himalayas or rejuvenate it.
- Increase production and planting of trees. These can be fuel, fodder, fruit and multi-purpose trees,
- Reduce or stop cutting of trees.
- Harvest water in small scale ponds and dams, roof rain-water harvesting etc.

- Rejuvenate springs wherever possible.
- Establish or promote local conservation and watershed management societies.
- Promote self reliance in Himalayan communities through small scale industries, whether agro based or non agro based. Examples of

Agro based	: Horticulture, Sericulture, Floriculture, Mushroom Culture, Apiculture, Medicinal Plants.
Non agro based	: Promote industry in mountain areas such as Electronics,and Optics, all of which are non-polluting and need quite a bit of indoor work and have light weight inputs as well as outputs.

- Entrust the rural people in the Himalayan communities after some training on tasks such as tailoring of clothes, khadi work as well as woollens to increase their level of income.
- Promote renewable energy, including possible establishment of small plants tapping renewable energy such as solar cookers, solar water heaters, solar driers, mini and micro hydel etc.

What Rotarians and other community leaders can do?

- Awareness building - Self education on Environment and Promotion of environmental awareness amongst all those within reach.

- "Begin not with the programme, but with a deed" says Florence Nightingale.
- Look out for waste material that can be utilized adding to the total value. Always direct your thinking in turning WASTE TO WEALTH. Designate community drop points for hazardous waste.
- Promote urban technology on waste utilization such as for solid waste matter. Promote both mechanical and semi mechanical plants for the same.
- Promote liquid sewage plants such as Okhla Sewage Plants with bio gas as byproduct which can be distributed to nearby colonies.
- Use of soiled night soil and liquid from slum areas in bio gas production programmes which can be provided to slum areas.
- Target elementary and secondary school students who will take ideas home. Present scholarships for environmental studies. Art from Garbage, tree nursery in each school and many others.
- Ten rules for success : The first is "Know how", the second is "Use what you know". The other eight are the same as the second.

Healing the Earth, Healing yourselves

The first principle of the right ecology of the mind is: Don's allow your mind to become a garbage disposal unit.

The second principle of the ecology of the mind is: No saving the Earth without changing our consciousness.

The third principle of the ecology of the mind calls on you to reflect for five minutes every day on who you are, what is your highest potential, what kind of world you would like to live in. After you have reflected, you must ask yourself how to act to bring about the kind of person you would like to be, to bring about the kind of world you would like to live in.

The ecology of the mind leads you to acquire ecological consciousness, which in turn leads you to right values, which are the foundations of right modes of action.

If you want to help the environment, help yourself become a gentle, unharmful person who interacts co-operatively and compassionately with all creation.

The ideas we have discussed here are a part of emerging Ecological Philosophy. What is ecological philosophy? Ecological Philosophy is to be friendly with nature and knowing why; is to be friendly with other forms of creation and knowing why; is to be friendly with your own body and knowing why; is to be friendly with your own mind and knowing why.

Ecological philosophy means declaring and maintaining peace with yourself and with all creation. Unless this is accomplished, we shall not save the Earth. Saving the earth is a spiritual crusade of getting your values and your philosophy right, so that you walk in peace because you are Walking Peace.

We say think and act environmentally while there is still time. In fact, time is running out on all of us. Floods, droughts, landslides, deforestation, also desertification are some of the evidences that are part of nature's warning.

We must heed these warnings. This reminds me of an incident when a person in a village which is getting flooded climbed up on his roof. A boat came to rescue him. He did not leave on the boat. "I have faith in God. He will take care of me. I will not go in your boat to a safe area." As the flood waters rose still more, a motor boat came to rescue him. He refused help with the same words. Finally when just his head was above the water, a helicopter came to rescue him. He still refused to be rescued. He died, and then complained to God. "My Lord, I had confidence that you will do something to save me. I was betrayed". The Lord showed his surprise and anguish and said, "Three times I sent means to save you. Even a helicopter. You refused every time. You did not even help even yourself in accepting a way to safety.

Time is short, heed the warning, both think and act environmentally while there is still time.

M.L.DEWAN.

Feedback and comments are welcome. Please write to :

Ravi Gulati
13, Khan Market,
New Delhi-110003 (India)
e-mail : mldewan@ndb.vsnl.net.in

Recommended Further Reading

The number of environmentally oriented publications has grown enormously in the past two decades. Since I haven't been able to read them all, I cannot be sure which are best. I can be sure that the following will provide useful information if you choose to dig deeper.

General Texts

G. Tyler Miller, *Living in the Environment : Principles, Connections and Solutions,* Belmont, CA, Wadsworth, 1994. (Get the 8th edition or newer.)

Daniel Chiras, *Environmental Science : Action for a Sustainable Society,* Redwood City, CA, Benjamin Cummings, 4th ed., 1994.

Cosmic Evolution

Duane Elgin, *Awakening Earth: Exploring the Evolution of Human Culture and Consciousness,* New York, Morrow, 1993.

Brian Swimme and Thomas Berry, *The Universe Story,* San Francisco, Harper, Allyn paperback, ed., 1994.

Cultural Evolution

Andrew Bard Schmookler, *The Parable of the Tribes : The Problem of Power in Social Evolution,* 2nd. ed., Albany, NY, SUNY Press, 1995.

Biological Evolution

Lynn Margulis and Dorian Sagan, *Microcosmos: Four Billion Years of Evolution From Our Microbial Ancestors,* New York, Summit Books, 1986.

Sustainable Society

Lester Milbrath, *Envisioning a Sustainable Society: Learning Our Way Out,* Albany, NY, SUNY Press, 1989.

Dennis Pirages, ed., *Footsteps to Sustainability,* Armonk, NY, M.E. Sharpe, 1996.

Sustainable Development

Dan Sitarz, *Agenda 21 : The Earth Summit Strategy to Save Our Planet,* Boulder, CO, Earth Press, 1993.

International Union for the Conservation of Nature and Natural Resources, *Caring for the Earth : A Strategy for Sustainable Living,* Gland Switzerland, 1991 (in collaboration with the United Nations Environment Program and the World Wide Fund).

Vandana Shiva, ed. *Close to Home : Women Reconnect Ecology, Health and Development,* Philadelphia, New Society Publishers, 1994.

Peter Bartelmus, *Environment, Growth, and Development : The Concepts and Strategies of Sustainability,* New York, Routledge, 1994.

Jeremy Carew-Reid, Robert Prescott Allen, Stephen Bass, and Barry Dalal-Clayton, *Strategies for National Sustainable Development : A Handbook for Their Planning and Implementation,* London, Earthscan Publications, 1994.

Ernst U. Von Weizsacker and Jochen Jesinghaus, *Ecological*

Tax Reform, A Policy Proposal for Sustainable Development, London and Atlantic Highlands, NJ, Zed Books, 1992.

Thaddeus C. Trzyna, ed., A *Sustainable World : Defining and Measuring Sustainable Development,* International Centre for the Environment and Public Policy, P.O. Box 189040, Sacramento CA 95818, USA, 1995.

Environmental Economics

Herman Daly and John Cobb, *For the Common Good: Redirecting the Economy Toward Community, the Environment, and a Sustainable Future,* Boston Beacon Press, 1989, 2nd. ed., revised and updated, 1994.

Ernst F. Schumacher, *Small is Beautiful: Economics As If People Mattered,* New York, Harper & Row, 1973.

Beat Burgenmeier, *Economy, Environment, and Technology: A Socioeconomic Approach, Armonk,* NY, M.E. Sharpe, 1994.

James Robertson, *Future Wealth, A New Economics for the 21st Century,* London, Cassell Publishers Ltd., 1989, paperback ed., 1990.

Ideas for Living Lightly on the Earth

Joe Dominguez and Vicki Robin, *Your Money or Your Life: Transforming Your Relationship with Money and Achieving Financial Independence,* New York, Penguin Books, 1992.

Duane Elgin, *Voluntary Simplicity : Toward a Way of Life That is Outwardly Simple, Inwardly Rich,* New York, William Morrow, 1981, 2nd. ed., 1993.

The New Physics

Fritjof Capra, *The. Turning Point: Science, Society and the Rising Culture,* New York, Simon & Schuster, 1982.

Murray Gell-Mann, *The Quark and the Jaguar: Adventures in the Simple and the Complex,* New York, W.H. Freeman, 1994.

The Biological Basis of Society

Mary Clark, *Ariadne's Thread: The Search for New Modes of Thinking,* New York, St. Martin's Press, 1989.

Anne H. Ehrlich and Paul R. Ehrlich, *Earth,* New York, F. Watts, 1987.

Paul R. Ehrlich and Anne H. Ehrlich, *Healing the Planet: Strategies for Resolving the Environmental Crisis,* Reading, MA, Addison-Wesley, 1991.

Environmental Education

David Orr, *Ecological Literacy : Education and the Transition to a Postmodern World,* Albany, NY, SUNY Press, 1992.

David Orr, *Earth in Mind : On Education, Environment, and the Human Mind,* Washington DC, Island Press, 1994.

Gregory A. Smith, *Education and the Environment : Learning to Live with Limits,* Albany, NY, SUNY Press, 1992.

Environmental Sociology

Charles Harper, *Environment and Society : Human Perspectives on Environment Issues,* Englewood Cliffs, NJ, Prentice Hall.

William Bullard, *Dumping in Dixie : Race, Class, and Environmental Quality,* Boulder, CO, Westview Press, 1990, 1994.

Environmental Psychology

Gerald T. Gardner and Paul C. Stern, *Environmental Problems and Human Behavior,* Needham Heights, MA,

Allyn & Bacon.

Theodore Roszak, Mary E. Gomes, & Allen D. Kanner, eds., *Ecopsychology: Restoring the Earth, Healing the Mind,* San Francisco, Sierra Club Books, 1995.

Environmental History

Donald Worster, *Nature's Economy : The History of Ecological Ideas,* New York, Cambridge University Press, 1985.

________ed. *The Ends of the Earth : Perspectives on Modern Environmental History,* New York, Cambridge University Press, 1988.

William Cronon, *Changes in the Land: Indians, Colonists, and the Ecology of New England,* New York, Hill and Wang, 1983.

Environmental Philosophy

Peter Wenz, *Environmental Justice,* Albany, NY, SUNY Press, 1988.

Warwick Fox, *Toward a Transpersonal Ecology,* Boston, Shambala, 1990.

Arne Naess (translated and edited by David Rothenberg), *Ecology, Community, and Lifestyle : Outline of Ecosophy,* New York, Cambridge University Press, 1989.

Environmental Ethics

Van Rensselaer Potter, *Global Bioethics : Building on the Leopold Legacy,* East Lansing, MI, Michigan State University Press, 1988.

Holmes Rolston III, *Environmental Ethics: Duties and Values in the Natural World,* Philadelphia, Temple University Press, 1988.

Environmental Political Philosophy

Robyn Eckersley, *Environmentalism and Political Theory : Toward an Ecocentric Approach,* Albany, NY, SUNY Press, 1992.

Robert Paehlke, *Environmentalism and the Future of Progressive Politics,* New Haven, CT, Yale University Press, 1989.

American Environmental Politics

Walter Rosenbaum, *Environmental Politics and Policy,* 3rd. ed., Washington, DC, Congressional Quarterly Press, 1995.

Norman Vig and Michael Kraft, eds., *Environmental Policy in the 1990s,* 2nd. ed., Washington, DC, Congressional Quarterly Press, 1994.

Global Environmental Politics

Garrett Porter and Janet Welsh Brown, *Global Environmental Politics,* Boulder, CO, Westview, 1995.

Sheldon Kamienecki, ed., *Environmental Politics in the International Arena : Movements, Parties, Organizations and Policy,* Albany, NY, SUNY Press, 1993.

Index of Key Concepts

Most concepts that are key to thinking environmentally are defined in the text and usually highlighted in bold letters. This index will help you to find them for a quick review. If a page number is highlighted in bold, that page has the main definition of the concept; other page numbers show usage of the concept.

Concept
1185/3/5-12-98 4 jay